Eileen Chevalier's

TYPEWRITER Aalphabet BOTANY

Teacher's Guide

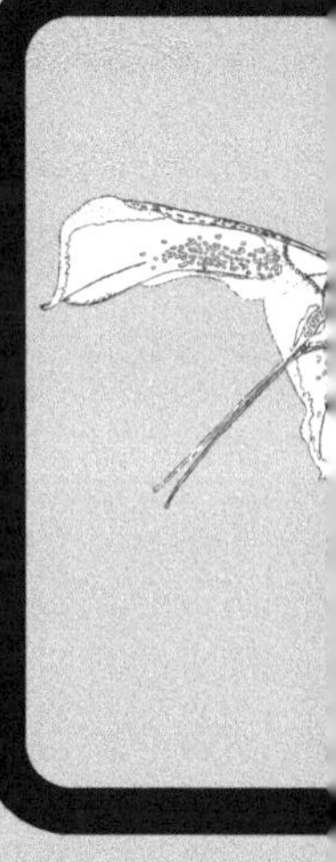

KNIGHT KING PRESS
Siler City, NC

written & illustrated by **Eileen Chevalier**

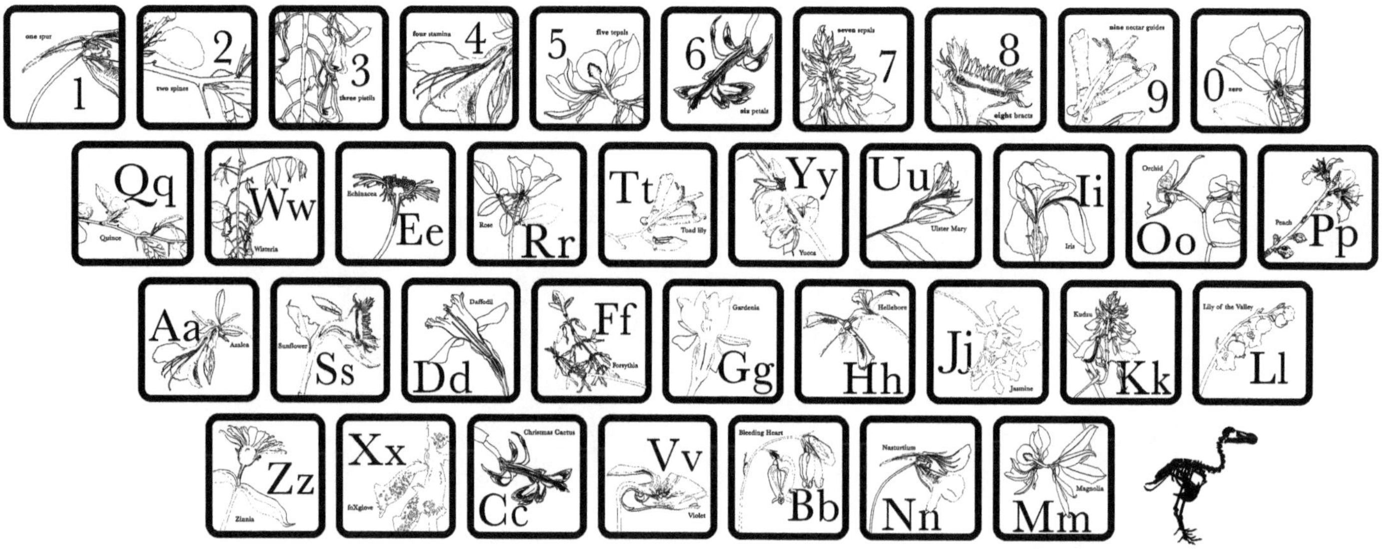

Knight King Press

ISBN 978-1-952704-13-0 (paperback) | ISBN 978-1-952704-26-0 (ebook)

Title: Eileen Chevalier's Typewriter Aalphabet Botany Teacher's Guide / Eileen Chevalier; illustrated by Eileen Chevalier.
Names: Eileen Chevalier, author | Eileen Chevalier, illustrator. | Eileen Chevalier, designer.
Subjects: LCSH Botany | Plants—Anatomy | Flowers—Pictoral works
Classification: LCC QK49 .C48 2026 | DDC 580 C527 tg 2026

Also in the *Eileen Chevalier's Typewriter Aalphabet* Series:
Botany Book: ISBN 978-1-952704-08-6 (hardback)
ISBN 978-1-952704-16-1 (paperback)
ISBN978-1-952704-17-8 (ebook)
Color + Label: ISBN 978-1-952704-14-7
Workbook: ISBN 978-1-952704-15-4

The artwork in this book was hand drawn by Eileen Chevalier.
The text is set in Century Gothic, Avenir Next, and Baskerville.

www.knightkingpress.com

KNIGHT KING PRESS

Eileen Chevalier's

TYPEWRITER Aalphabet BOTANY

Welcome to Botany!

This Teacher's Guide is designed to be used with Eileen Chevalier's Typewriter Aalphabet Botany Curriculum. The goal of this program is to enable students (and you!) to discover the beauty of plants, wrestle with the complexity of living systems, and develop critical thinking.

This study is different from others because most of the questions don't have simple answers. Some of them can't even be answered with the information that is now available. That's good news! The best questions are the ones that make you dive deeply into a subject, leading you to even more mysteries. In fact, many of the things scientists once believed to be true were later disproven. Even the classification system is in flux. While this may seem disheartening at first, it can actually make the study of science more exciting. There is always more to discover!

Here are some great phrases to adopt:

That's how we understand it right now. Let's keep an open mind as we do more research.

How could we design an experiment to test it? Can we know for sure? Why or why not?

Great observations! What else might be happening that you aren't able to see?

What other details and labels would add to your drawing so that you can reference it later?

I don't know. What would be the best way for us to learn more? Let's learn together.

You can't know it all, but you can model curiosity and the pursuit of understanding, not just knowledge. Focus on broad concepts and meaning instead of rote, trivial facts. If you and your students are exploring and attempting and making mistakes, celebrate! All of those things are part of learning. Your efforts will be rewarded if you work with an inquisitive and open spirit.

Keep exploring, discovering, and finding hidden beauty!

Eileen Chevalier

Thank you for purchasing the Botany Teacher's Guide!
As an artist, teacher, and researcher, I am thrilled to have you on this learning journey with me. My sincere hope is that you, too, have created (or are in the process of creating) something to share with the world. Let's respect one another so that we are each encouraged to continue making the world a more wonderful place in which to learn, grow, and create.

All the best!
Eileen

Table of Contents

Eileen Chevalier's
TYPEWRITER Aalphabet BOTANY

There are many components of the Botany Curriculum, including a library of Lesson Videos. The goal is to provide a flexible system that can be adapted to a variety of settings and learners. Please feel free to use the materials that work best for you.

Eileen Chevalier's Typewriter Aalphabet Botany Curriculum is an exploration of botany - both the natural world and human attempts to categorize and conceptualize it. Because life is a beautiful, complex playground in which to explore, the Botany Curriculum emphasizes the rich diversity of nature and our inadequate attempts to quantify it.

Learning happens when we physically and intellectually engage with material in challenging ways that facilitate connections and trust in our own ability to discover and make sense of the world around us. Most questions do not have simple answers. The questions presented in this guide are crafted to develop logical thinking skills and creative problem solving while effectively teaching botanical concepts.. The activities provide opportunities for tangible interactions with plants and categorization systems.

The Botany Curriculum is not just about plants. It is for the development of human potential. This Teacher Guide provides teachers with tools for developing themselves as well as their students. We are all on the learning journey together!

Classroom Resources:

Eileen Chevalier's Typewriter Aalpabet ***Flashcards***
These 36 cards contain full-color illustrations of each of the letters and numbers; information on the backs is used throughout the lessons. Cards can be sorted as concepts are learned, such as organizing them by clade, by annual or perennial, by edible or poisonous, etc. In addition, you can hang the cards as a QWERTY keyboard on your classroom wall.

Eileen Chevalier's Typewriter Aalphabet ***Botany Book***
The *Botany Book* combines 8-inch versions of the 36 Flashcards (front and back) with the definitions and labels of the Color + Label Book, the Clades Poster, and select reproductions of Eileen's notes and drawings from her personal set of 6" Art Cards. Also includes an index and four-page glossary with several color illustrations. The *Botany Book* is an engaging reference book on its own or as part of the Botany Curriculum.

Consumable Student Resources:

Eileen Chevalier's Typewriter Aalpabet ***Botany Workbook***
Students can record all the definitions, short-answer questions, research prompts, and activities in the Botany Curriculum. Each lesson perfectly aligns with the Teacher Guide and lesson videos.

Eileen Chevalier's Typewriter Aalphabet ***Botany Color + Label Book***
Students can record notes, add drawings to the glossary/index, reference the labels to understand plant parts, and color the illustrations.

Botany Videos

Lesson Videos

There is an extensive library of free lesson videos for you to use with this curriculum. Please visit www.knightkingpress.com/botany for the links. Feel free to play (and pause as needed) the videos for your class and/or use them as a resource as you prepare to teach the lessons.

Lily Dissection Video

The lily dissection video offers a clear explanation of basic monocot flowers as Eileen dissects a lily and completes the Lily Dissection Worksheet. Each student will need a copy of the Lily Dissection Worksheet on pages 9-10 (print double-sided and fold). Ideally, students would be able to dissect their own lily along with the video. If flowers are not available, students can watch and complete the worksheet based on the images in the video. If you conduct the dissection without the video, please preview it for the correct terminology.

Dissection Videos

Many of the dissection videos are of flowers that are in the Typewriter Aalphabet. Print out corresponding coloring pages for each student to label and color as they watch the video.

Botany Books Used in this Curriculum

The following books are used as reference books for this curriculum. Each is fantastic and worth purchasing if funds are available. I also highlight specific references in the lesson videos. Although they are listed alphabetically below, the highest priority for the curriculum are: *Flora (DK), Natural History (DK Smithsonian)*, and the *Ultimate Visual Dictionary (DK).*

Of course, the Flashcards are a fantastic addition! If you prefer a bound book to the loose-leaf cards, Eileen Chevalier's Typewriter Aalphabet Botany Book is now available.

Picture books (great for all ages): *A Seed is Sleepy, Gregor Mendel*, and *Flowers are Calling*. Very young or sensitive students may not be ready for: *Wicked Plants*.

Aston, Diana Hutts. ***A Seed is Sleepy***. Chronicle Books, 2007.

Bardoe, Cheryl. ***Gregor Mendel: The Friar Who Grew Peas***. Abrams Books for Young Readers, 2006.

Chevalier, Eileen. ***Eileen Chevalier's Typewriter Aalphabet Flashcards***. Knight King Press, 2020.

Flora: Inside the Secret World of Plants. DK, 2018.

Gray, Rita. ***Flowers are Calling***. Houghton Mifflin Harcourt, 2015.

Illustrated Dictionary of Biology. Usborne, 2007.

Miessler, Diane. ***Grow Your Soil!*** Storey Publishing, 2020.

Natural History. DK Smithsonian, 2021.

Sanchez, Anita. ***Karl, Get Out of the Garden!*** Charlesbridge Publishing, 2017.

Stewart, Amy. ***Wicked Plants: The Weed that Killed Lincoln's Mother & Other Botanical Atrocities***. Algonquin Books of Chapel Hill, 2009.

Ultimate Visual Dictionary. DK, 2011

Scheduling the Lessons

YEAR-LONG SCIENCE:
If this will be your curriculum for the year, you can complete one lesson each week for a 36-week school calendar. Your weekly schedule can be divided into two days: lesson, definitions, and questions on day one; activities and journaling on day two. If you have three days, add another day for journal activities or add in a dissection video.

To accomplish writing and public speaking objectives, have students polish their answers into papers and present them to the class. The poster activity in Lesson 29 would be a great alternative for students who excel with visual communication.

SHORTENED YEAR/SUPPLEMENT:
To shorten the time required, choose the most important lessons to meet your goals or focus on specific parts of each lesson. For instance, you can just complete the *Roots, Stems*, and *Leaves* lessons and skip the *Not Roots, More Stems*, and *More Leaves*.

To cover more lessons in less time, watch a lesson, complete the definitions, and answer just one question of your choosing. If students are in groups, they could each complete a different question or activity and then present their ideas and findings to the class.

As a supplement to high school level biology, students should complete as many lessons as possible. Advanced students can accurately document experiments and type longer form answers with quotes, notations, and bibliographic informtion. Focus on *Plant Systems* lessons.

CLASS SCIENCE FAIR:
Choose one lesson that will relate to an engaging science fair project idea (photosynthesis in leavs indoors vs. outdoors; propagation with various plant parts and/or conditions; etc.). Complete the lesson and have students brainstorm quesions they want to answer and methodologies they could use in their experiments.

ACTIVITY FOCUSED:
If you are most interested in having students working directly with plants, you can go though the lesson and move directly into the activities. Choose lessons with the activities you are most equipped to complete. Almost every lesson has more than one activity. Pick the one you like best, do each on the same day, or take multiple days to complete the activities. There are a few activities that take longer (growing avocado, letting ginger dissentigrate), so they can be schedule to begin before the actual lesson.

BLOOM TIME
Some activiites will be easier to complete during specific seasons. As they vary widely depending on your location, plan ahead to take advantage of your growing conditions.

ACTIVATION
Plants are inspiring. Try bringing in a plant or plant part to activate the students' interest. If you can't find fresh flowers or branches, don't worry. Much of the food we eat is from plants. Ask students which part of the plant your coffee comes from before the *Fruit* lesson or have them guess the most expensive plant part (saffron) before the *Pistil & Stamen* lesson.

MOST IMPORTANTLY, trust yourself as a teacher and adjust for your own students' needs.

Lily Dissection

**Lilies are monocots,
part of the monocotyledoneae clade.**

See if you can find some common characteristics of monocots in this lily.

1. tepals instead of petals or sepals
2. tepals in groups of three
3. leaves with:

- long, parallel veins (parallel venation)
- smooth edges (entire margin)
- long, thin shape (linear shape)

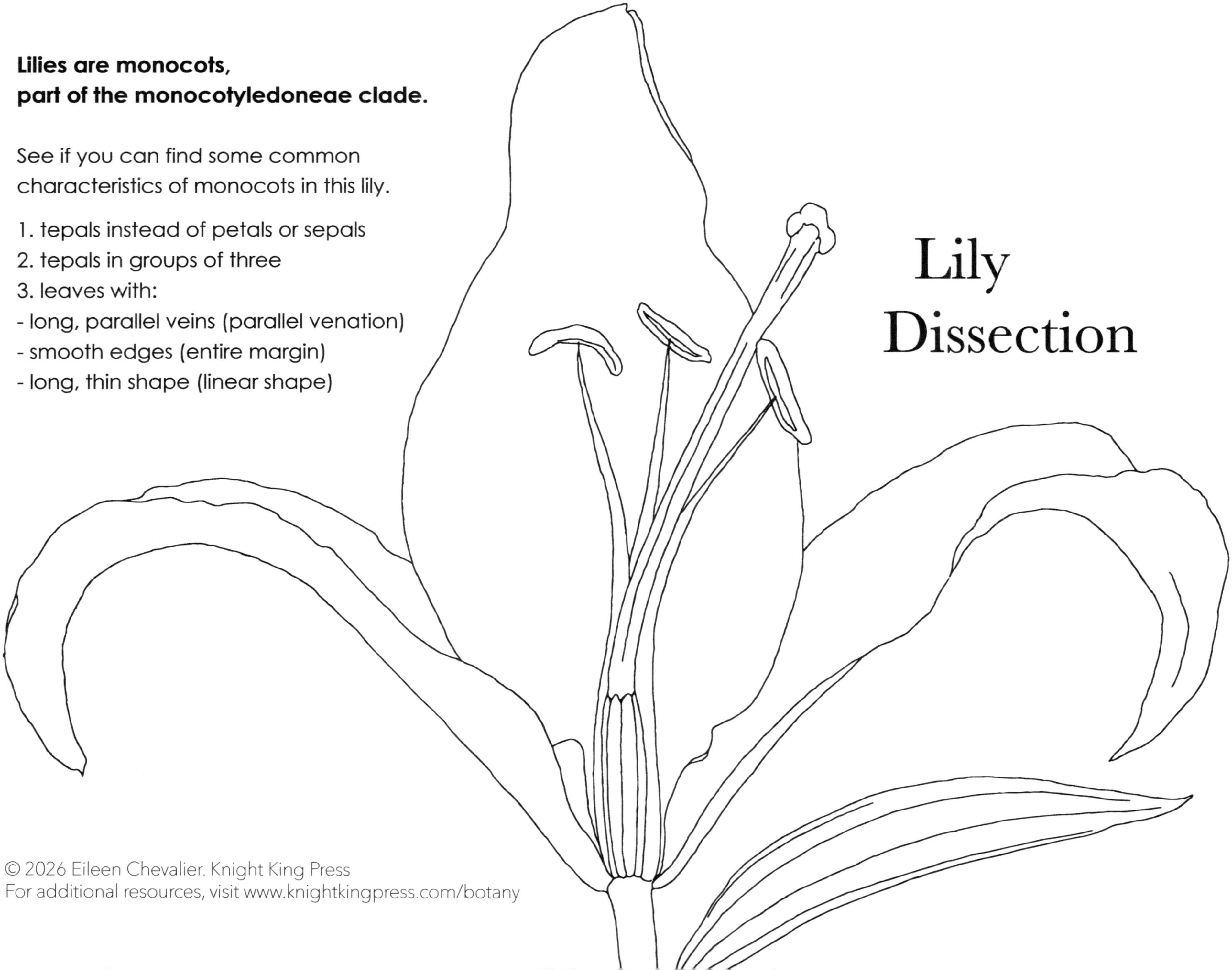

Petals- showy parts around the stamens, usually brightly colored.
Sepals- leafy parts around the petals, usually small and green.
Tepals- name for all petals and sepals if they look very similar.

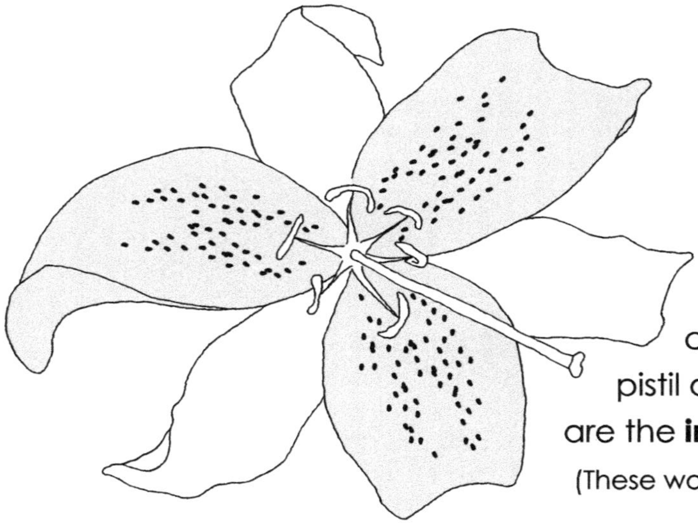

The 3 tepals closest to the pistil and stamens are the **inner tepals**.
(These would be petals.)

Monocots, like the lily, usually have **tepals** instead of distinct petals and sepals.

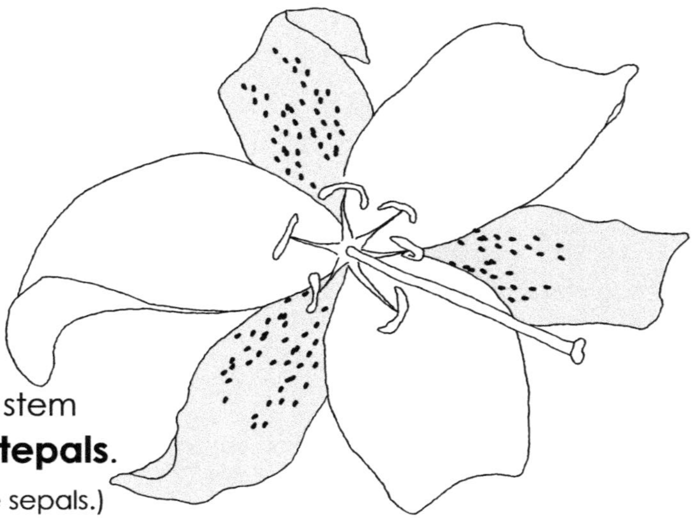

The 3 tepals closest to the stem are the **outer tepals**.
(These would be sepals.)

All the tepals (petals and sepals) of a flower are collectively called the **perianth** because they are around (peri-) the <u>anth</u>ers.

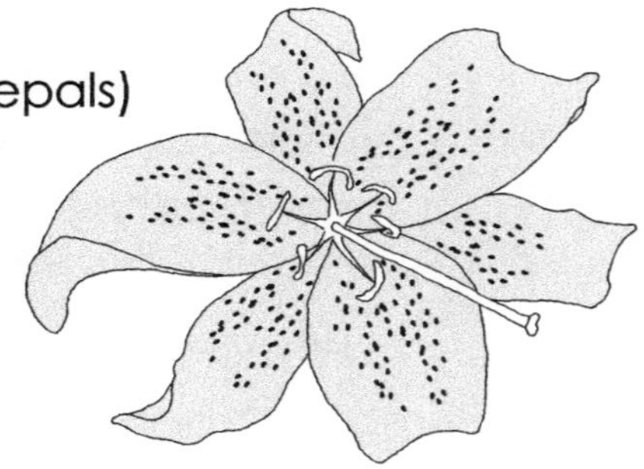

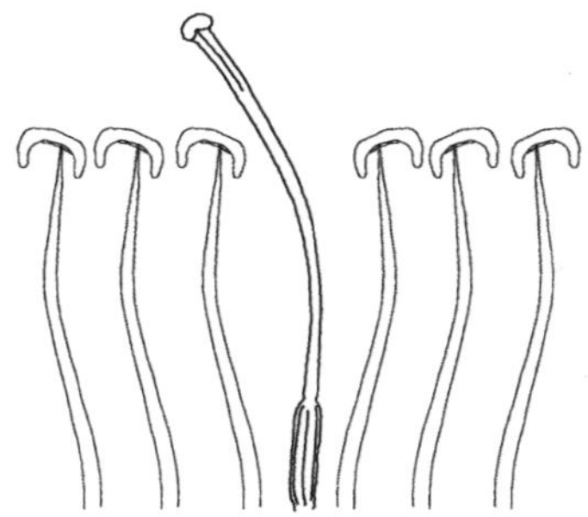

The lily has 1 **pistil** surrounded by 6 **stamens** (or stamina). Do you see the pistil in the center of the six stamens in the drawing above?

A **stamen** has pollen on its **anther**.

The anther is at the top of the stamen's long **filament**.

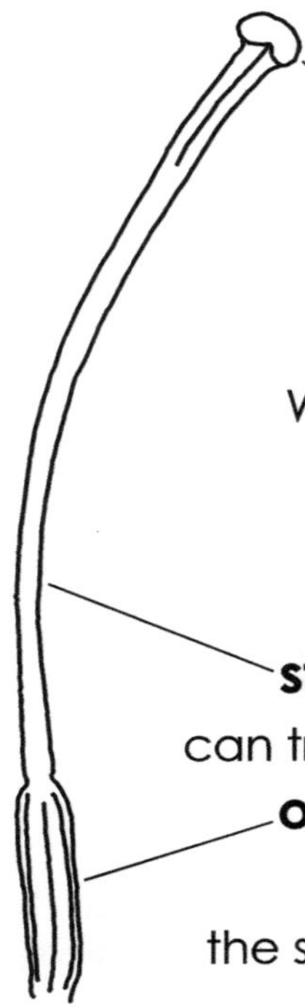

A **pistil** has a sticky **stigma** at the top for catching pollen grains.

When pollen lands on the stigma, a tube is formed through the long **style** so the pollen can travel down to the **ovary** at the base of the pistil where the seeds are formed.

INTRODUCTION: WHAT IS BOTANY?

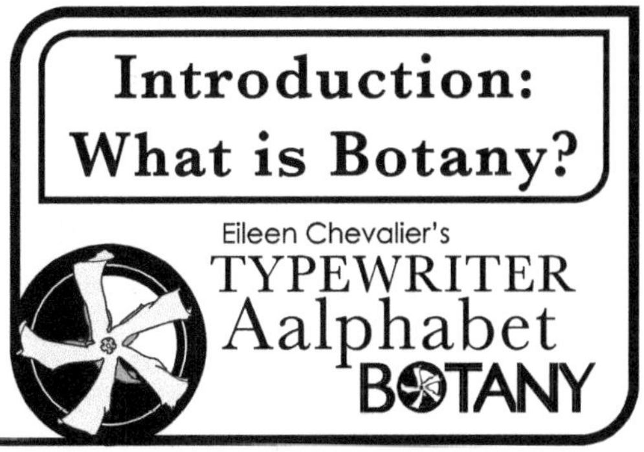

Welcome to Eileen Chevalier's Typewriter Aalphabet Botany Curriculum. You can watch the lesson video online for an introdcution to the program and the Flashcards. Enjoy!

www.knightkingpress.com/botany

NOTES

READ

Eileen Chevalier's Typewriter Aalphabet FLASHCARDS

DEFINE

botany - the scientific study of plants
scientific - using the methods and principles of science
Questions: Have the methods and/or principles of science changed? Why? Who invented the Scientific Method? When?
organism - a living thing made up of interdependent parts
Questions: What is an example of an organism? What isn't?
interdependent - two or more things that are distinct in form and function yet dependent on one another for their existence
Questions: What would happen to fruits and vegetables if there were no bees? What would happen to bees if there were no flowers?

QUESTION & INVESTIGATE

Look through the cards and think about how many different ways you could organize them. Which method might be the most helpful? How does your goal determine which way to organize them?

JOURNAL or BOTANY WORKBOOK

Copy an illustration or do an original drawing while observing a plant. Don't worry if your drawing isn't accurate. The goal is to observe closely. You may even want to try a blind drawing: draw each curve on your paper as you follow it with your eyes. Only look at the flower. Do not look at your paper until your drawing is done.

Make your own alphabet card. You can use the same format as the Typewriter Aalphabet cards or create your own. Do you want to add fun facts?

SCIENTIFIC CLASSIFICATION

In this lesson, we'll learn about scientific classification system, which is a common language for organizing all living things. Like all classification systems, it's not perfect. Like all languages, it continues to evolve.

www.knightkingpress.com/botany

Scientific Classification

Eileen Chevalier's
TYPEWRITER
Aalphabet
BOTANY

NOTES

READ

Karl, Get out of the Garden! Anita Sanchez
Gregor Mendel: The Friar Who Grew Peas, Cheryl Bardoe
Natural History, DK Smithsonian (28-29, 32-33, 549)

DEFINE

binomial nomenclature - a two-part naming system that Carl Linnaeus developed in the 1700s. Genus is capitalized. Species is lower case. Both are italicized: *Genus species*

> What is the binomial nomenclature for human? What other species share our genus? What does sapien mean? Is this accure? Why or why not?

QUESTION & INVESTIGATE

Latin and Greek are often used in the classification system. Now we mix those with English endings: stamens instead of stamina. How else has it changed? Why? Will it continue to change? Why?

Look at the scientific names on each Flashcard. Can you tell which come from Latin or Greek? Which are named for people?

Why does Forsythia (named after William Forsyth, a founding member of the Royal Horticultural Society in England) have an x between the genus and species?

JOURNAL or BOTANY WORKBOOK

Find an interesting binomial nomenclature, look up its etymology, draw the plant, and explain why the name intrigues you.

Draw a plant - real or imaginary - and make up your own binomial nomenclature for it.

Make up a binomial nomenclatures for your family members. Do you all have the same genus? What are the different species? What makes each person unique and wonderful?

COMMON NAMES

Each plant species has one scientific name, but can have many common names, especially if the plant grows in many different places or is used for different purposes. In this lesson, we will discuss common names and compare them with scientific names.
www.knightkingpress.com/botany

Common Names

Eileen Chevalier's TYPEWRITER Aalphabet BOTANY

NOTES

READ

Karl, Get out of the Garden! Anita Sanchez

DEFINE

common name - informal name that is widely used, often describing the physical or medicinal properties of the plant.

Why is the bleeding heart sometimes called "lady in a bath?"

QUESTION & INVESTIGATE

Look at the common names for the plants used in medicine on page 10 of *Karl, Get out of the Garden!* What do you think these plants might have been used for? How might these common names be more useful than the scientific names?

What were the common names for dandelion? Why do you think it had so many?

Compare the common names with the binomial nomenclature names on the cards. What do you notice? What is similar or different? Does the letter on the card match the common name or the scientific name?

Compare *Dicentra formosa* (Bb card) with *Dicentra spectabilis* in the dissection video. *Dicentra spectabilis* has a new genus: *Lamprocapnos*. Why? What is the effect of a classification change?

JOURNAL or BOTANY WORKBOOK

Find an interesting common name, draw the plant it represents, and explain why the name intrigues you.

Draw a plant - real or imaginary - and make up your own common name for it.

Alstromeria angustifolia has several common names: Peruvian lily, lily of the Incas, parrot lily, or Ulster Mary. Why do you think *Alstromeria* has those names? Make notes for yourself (and drawings if possible) about where these names come from.

PLANTAE DIVISIONS

All the plants in the Typewriter Aalphabet are in the division Angiospermae. In this lesson, we'll find out how to tell if a plant is part of the Angiospermae division and learn more about the plants that aren't.

www.knightkingpress.com/botany

Plantae Divisions

Eileen Chevalier's TYPEWRITER Aalphabet BOTANY

NOTES

READ

Natural History, DK Smithsonian (106-207)

DEFINE

angiospermae - division of Plantae kingdom containing all flowering plants
angiospermophyta - phylum of the Plantae kingdom containing all flowering plants

What is a division?
How does it relate to a phylum?

QUESTION & INVESTIGATE

Angiospermae means covered (*angio*) seeds (*sperm*). Is this an easy way to identify Angiosperms? What seeds aren't covered? What plants create them? Where in the classification system would these plants be found?

How can plants reproduce without seeds? Which plants are able to do that? Are they all part of the same division?

JOURNAL or BOTANY WORKBOOK

Find examples of other divisions of the Plantae Kingdom. Draw from real life if possible or use pictures to draw and label examples in your journal. Make sure to include which division each plant is in. Bonus: try to include how the plant reproduces.

Angiosperms are extremely diverse. They range from giant trees to tender plants, from water-dwellers to those in deserts, and their flowers can be tiny or enormous. Explore the diversity of the Angiospermae Division and illustrate your findings. You can use the Aalphabet cards. Magnolia is a tree. Bleeding hearts die back completely each year. Jasmine and wisteria are long vines. Sunflowers can grow taller than humans, but violets bloom just above the ground.

ANGIOSPERMAE: BASAL & CLADES

In this lesson, we'll look at the classifications of the Angiospermae Division. In addition to the Basal Angiosperms, there are three Clades: Magnoliidae, Monocotyledoneae, and Eudicotyledoneae.

www.knightkingpress.com/botany

Angiospermae: Basal & Clades

Eileen Chevalier's TYPEWRITER Aalphabet BOTANY

NOTES

READ

Natural History, DK Smithsonian (122-207)

DEFINE

clade - a branch in the tree of life; in evolutionary terms, all members of a clade are descended from a common ancestor

> Why are there three (not two or four) clades? Are there angiosperms that aren't part of the three clades?

QUESTION & INVESTIGATE

Angiosperms were once classified as either dicot or monocot, but are now split into four groups: Basal Angiosperms, Magnoliids, Monocots, and Eudicots. Does this division make sense? Why might Angiospermae be divided into even more groups?

Clade comes from the Greek word *klados*, meaning *branch*. Clades can occur at any level of the taxonomic rank, and are not part of the original taxonomic system. However, the term clade is often used instead of class in the Plantae Division. Why?

Sort the Flashcards into Eudicots, Monocots, and Magnoliidae. Is the number of each representative of their group size?

JOURNAL or BOTANY WORKBOOK

Compare the plant parts of the three clades. How are the roots, stems, leaves, and flowers different? Draw examples of each and take notes to compare.

Collect examples of each group (basal angiosperms, magnoliidae, monocots, and eudicots). Try drawing, taking pictures, doing a rubbing, or pressing them. Label with the group and where you found the plant, as well as the scientific and common names.

Which group of plants do we eat most often? Document what you have for one meal and which clades the ingredients are from. (Don't forget spices and beverages!)

MONOCOTYLEDONEAE CLADE

Monocotyledoneae were one of the two original groups of Angiospermae. In this lesson, we'll look at the roots, stems, leaves, and flowers of Monocots to discover their distinguishing characteristics.

www.knightkingpress.com/botany

Monocotyledoneae Clade

Eileen Chevalier's TYPEWRITER Aalphabet BOTANY

NOTES

READ

Natural History, DK Smithsonian (130-149)
Ultimate Visual Dictionary, DK (126-127)
Flora: Inside the Secret World of Plants, DK (60-61)

DEFINE

monocotyledoneae - clade (branch) of the Angiospermae Division that contains plants with one (mono) seed leaf (cotyledon)

> Can you identify a monocot if you can't see the seed?
> Can you be sure?

QUESTION & INVESTIGATE

Look through DK *Natural History* for an overview of how diverse the Monocotyledoneae clade is. How diverse are the sizes and shapes of the plants and their plant parts (roots, stems, leaves, flowers)? How diverse are their ecosystems?

Examine monocot trees. What do you notice about the structure of their trunks? Examine the micrograph of the monocot stem in *Flora*. How might this inner structure affect the shape of the tree?

Examine monocot flowers. Can you identify the petals and sepals? Read the Ll and 5 cards to find another way to name them. How many petals/sepals/tepals are on each flower. Can they all be divided by the same number?

JOURNAL or BOTANY WORKBOOK

Examine, copy, and label the micrograph of the monocot stem in *Flora*. Examine, copy, and label the micrograph of the monocot root in *Ultimate Visual Dictionary*. Compare.

Gather many different types of monocot leaves to study and compare. Use a loupe to see details.
Venation (how veins branch): Draw or do a rubbing of the venation patterns.
Shape: What do the **silhouettes** look like? Trace the outer edge to see the shape.
Margin: How does the **edge** look and feel? Draw a close-up of the margin.

EUDICOTYLEDONEAE CLADE

The Eudicotyledoneae Clade is the largest group of Angiospermae. (38 orders, 307 families, 182,227+ species) In this lesson, we'll look at their roots, stems, leaves, and flowers to find their distinguishing characteristics.

www.knightkingpress.com/botany

Eudicotyledoneae Clade

Eileen Chevalier's TYPEWRITER Aalphabet BOTANY

NOTES

READ

Natural History, DK Smithsonian (150-207)
Ultimate Visual Dictionary, DK (126-127)
Flora: Inside the Secret World of Plants, DK (60-61)

DEFINE

dicotyledoneae - archaic classification for a group of plants that produce seeds with two cotyledons (seed leaves)
eudicotyledoneae - a clade (branch) of flowering plants that produce seeds with two cotyledons (seed leaves)

> Why is the dicot/monocot system archaic (from a previous time) but not obselete (no longer used)?

QUESTION & INVESTIGATE

Consider the difference between dicots and eudicots? Read your Nn card. What does the prefix "eu-" mean? Why is it important to distinguish between dicots and eudicots?

A "rosate" corolla (all the petals of one flower) has five petals that usually overlap. Look at the petals of three common "berries": strawberry, blackberry, and blueberry. Which orders do these plants belong to? What is the corolla clue? Is the number five important?

JOURNAL or BOTANY WORKBOOK

Examine, copy, and label the micrograph of the eudicot stem in *Flora*. Examine, copy, and label the micrograph of the dicot root in *Ultimate Visual Dictionary*. Compare. How do they relate to the monocot stem and root micrographs?

Gather many different types of eudicot leaves to study and compare. Use a loupe to see details.
Venation (how veins branch): Draw or do a rubbing of the venation patterns.
Shape: What do the **silhouettes** look like? Trace the outer edge to see the shape.
Margin: How does the **edge** look and feel? Draw a close-up of the margin.

MAGNOLIIDAE CLADE

Small but diverse, the Magnoliidae include magnolia, black pepper, nutmeg, avocados, and cinnamon. In this lesson, we'll look at what makes a Magnoliidae special.

www.knightkingpress.com/botany

Magnoliidae Clade

Eileen Chevalier's TYPEWRITER Aalphabet BOTANY

NOTES

READ

Natural History, DK Smithsonian (128-129)
Flora: Inside the Secret World of Plants, DK (60-61)

DEFINE

magnoliidae - small, diverse clade of the Angiospermae Division containing plants that usually have pollen with a single pore and tepals like monocots, but often have two cotyledons and woody stems like the eudicots.

> The DK *Natural History* book states that magnoliids are "botanically between basal angiosperms and monocots." Why, then, do they share eudicot traits when eudicots are an entirely different branch?

QUESTION & INVESTIGATE

Compare the images of monocot and dicot stems in *Flora*. Then compare trees of the three clades in DK *Natural History*. Which type of stem might the Magnolia tree have?

Magnolia flowers are encased before opening, just like the eudicots. However, the outer structure is a bract instead of a sepal. Compare the emergence of new magnolia leaves with new magnolia buds. What is similar? Are the leaf coverings sepals or bracts? What happens to these plant parts after the leaves and flowers emerge?

Compare the cones of conifers to the seed pods of the magnolia. How are they similar or different? How does the corolla of a magnolia compare to eudicots and monocots?

JOURNAL or BOTANY WORKBOOK

You probably have many magnoliidae products in your house. Go on a hunt to find examples of magnoliid, monocot, and eudicot products. Document your findings!

Germinate avocado seeds and document the process. Create a log where you record how long you soak the pits, the day they split, the day a root first appears, and then record the root and stems lengths daily. You may want to set aside 2-3 pages.

GERMINATION

All Angiospermae can grow from the seeds that their flowers create. In this lesson, we'll use some different seeds to examine how the germination process works.

www.knightkingpress.com/botany

Germination

Eileen Chevalier's TYPEWRITER Aalphabet BOTANY

NOTES

READ

Grow Your Soil, Diane Messler
A Seed is Sleepy, Diana Hutts Aston & Sylvia Long
Ultimate Visual Dictionary, DK (152-153)
Flora: Inside the Secret World of Plants, DK (284-293)

DEFINE

embryo (germ) - organism in the stage of the life after fertilization and into germination. This stage begins when a zygote divides into two cells and continues until it forms cotyledons (seed leaves), hypocotyl (primary stem), and radicle (primary root).
germination - the process of a seed embryo developing into a seedling

How is the embryo protected as it develops?

QUESTION & INVESTIGATE

Embryos are just one part of the seed. All seeds are protected by a seed coat (testa) and some seeds, like corn, also contain an endosperm (endo=within; sperm=seed) that provides nutrition for the embryo. It can be delicious (this is why corn is juicy). Open fresh corn kernels to find the endosperm.

Germinating seeds need specific temperature, light, and soil texture. Some even need fire to get started! How do these adaptations help or hinder the plant's ability to survive? Can humans use these adaptations in the production and storage of seeds?

JOURNAL or BOTANY WORKBOOK

Read *Grow Your Soil* and learn about healthy soil. Experiment with starting seeds in various mixes - compost, regular dirt, store bought soils - and record your findings daily. Which works best? Why?

Copy the germination illustration in *Flora* (p. 286) or *A Seed is Sleepy* ("Generous") and label. *NOTE: The illustration in "A Seed is Generous" shows only one type; some seeds can have more than two cotyledons. Also, the plumule is the stem of the embryonic shoot, not just the leaves.

ROOTS

Root systems are just as diverse as stems and leaves. In this lesson, we'll look at roots, one of the three main plant parts. We'll compare fibrous roots and tap roots, find roots that store food, and check out other features.

www.knightkingpress.com/botany

Roots

Eileen Chevalier's TYPEWRITER Aalphabet BOTANY

NOTES

READ

Ultimate Visual Dictionary, DK (126-127, 132-133)
Flora: Inside the Secret World of Plants, DK (18-55)

DEFINE

root - part of the plant, usually underground, that collects and transports nutrients and water, anchors the plant, and often stores nutrients and food (carbohydrates).

What makes a root different from a stem or leaf?
Do all plants have roots?

QUESTION & INVESTIGATE

The *Ultimate Visual Dictionary* says that roots are "underground parts of plants," but is that always true? Can you think of roots that aren't underground? Can you think of leaves or stems (or even an orchid flower!) that are underground? (*Flora*, pages 38-9)

Roots usually send out tiny root hairs. How would this help the plant accomplish its three main jobs? What other advantages would the root hairs provide?

Compare the roots of monocots and dicots. In addition to the different structure within each root (*Visual Dictionary*, 126-7), the overall structure is also different. Compare the fibrous roots of monocots with the taproots of eudicots. (*Flora*, 38-9) How might each type be beneficial to the plant?

JOURNAL or BOTANY WORKBOOK

Dig up a small, deep area in the soil. Notice how the soil around the roots is different from the soil below. If possible, dig in soils with different structures: clay soil, sandy soil, garden soil. Are the roots the same in all three? Use pictures to record your findings.

Look at the roots of plants for yourself. Try buying a plant at a nursery, pulling up small weeds after a rain, or using a shovel to get deeper roots. Can you find the tap root or it is a fibrous root? How dry are they? How wide, long, and deep? If you were able to pull up more than one, how do they compare? Record your observations.

STEMS

Stems create structure for plants and are responsible for transporting water and nutrients. In this lesson, we'll look at various stems and compare their structure, vascular systems, and defense mechanisms.

www.knightkingpress.com/botany

Stems

Eileen Chevalier's TYPEWRITER Aalphabet BOTANY

NOTES

READ

Natural History, DK Smithsonian (150-207)
Flora: Inside the Secret World of Plants, DK (50-51, 56-107)

DEFINE

stem - part of the plant that provides structure and transports water, nutrients, and sugars up to the leaves and down to the roots; may also perform photosynthesis and store nutrients.

Does every plant have a stem?

QUESTION & INVESTIGATE

The Ultimate Visual Dictionary says that stems grow above ground. Is this always true? Where else can stems grow? (*Flora*, 50-51)

There are so many types of stems: vines, tree trunks, dainty petioles, herbaceous and woody, even Christmas cactus stems with no leaves. What are the advantages of each? How can each type use their stem to ensure survival of itself and its species?

Look up pictures of maple sap being harvested to create maple syrup. What is sap? Where is it coming from and where is it going? Look back at your drawings of stems to see what the taps are boring into. Why is it a super food? Finish your investigation with some pancakes and maple syrup!

JOURNAL or BOTANY WORKBOOK

Find a diagram of a stem, copy it into your journal and label it. Add your thoughts from the questions above and make up questions of your own!

Sort the cards into herbaceous or woody stems and make a list in your journal. Some are obvious and you'll learn about the others. When you encournter new plants, add them to your list.

People aren't the only creatures who want to tap into the sweet liquids running through plants. Draw and/or collect pictures of how stems protect themselves. (bark, thorns, etc.)

LEAVES

Leaves are often the most prominent part of the plant. In this lesson, we will learn how to identify various characteristics (shape, margin, venation, and arrangement) and look at some interesting leaves.

www.knightkingpress.com/botany

Leaves

Eileen Chevalier's TYPEWRITER Aalphabet BOTANY

NOTES

READ

Ultimate Visual Dictionary, DK (136-139)
Flora: Inside the Secret World of Plants, DK (108-181)

DEFINE

leaf - flat structures extending from the stem of the plant where photosynthesis takes place

Does photosynthesis happen anywhere else?
Does every plant have leaves?

QUESTION & INVESTIGATE

There are many ways to classify leaves: their shape (silhouette), margin (edge), venation (how the veins flow through the leaf), and arrangement (how leaves are attached to the stem). Which might be the most important aspect for identification?

How else can you identify/classify leaves? Which characteristics would be important? Which would be important to the plant? What do oxalis leave do at night? Why? Why are succulent leaves not flat?

Leaves are an important source of food on earth. Why? Where does the nutrition in leaves come from? Besides nutrients, what else do leaves contain for human health?

JOURNAL or BOTANY WORKBOOK

Find a diagram of a leaf (*Visual Dictionary*, 136). Copy it into your journal and label it. Add your thoughts from the questions above and make up questions of your own.

Find leaves that you can cut from their plant. (Be sure to watch out for compound leaves - get all the leaflets!) Do rubbings by placing the leaf under your paper and lightly rubbing a crayon or the side of your pencil over the leaf. Label the shape, margin, and venation using *Flora*. How many different types of leaves can you collect?

NOT ROOTS

Not everything that grows underground is a root. Many plants have special stems and leaves that are mistaken for roots. In this lesson, we'll take a closer look at bulbs and their spectacular roots.

www.knightkingpress.com/botany

Not Roots

Eileen Chevalier's TYPEWRITER Aalphabet BOTANY

NOTES

READ

Ultimate Visual Dictionary, DK (154-155)
Flora: Inside the Secret World of Plants, DK (86-87)

DEFINE

bulb - underground basal plate stems attached to scale leaves that store food for the plant

How does a bulb survive without roots?

QUESTION & INVESTIGATE

The basal plate stem is like a stiff pancake with scale leaves attached. When conditions are right, roots emerge from below while leaves and flower stems grow from the top. Cut open a bulb (ie-onion) and examine the basal plate stem. How is it similar to and different from other stems? Do basal plate stems have the same job as other stems?

How do most stems survive? How is the survival of bulbs different?

Read about contractile roots on pages 36 and 37 of *Flora*. What makes a contractile root special? Are all the roots of a bulb the contracting type? What are the advantages of having contractile roots? Try buying bulbs at a store and planting them in rocks in a glass container. Watch as the roots emerge from the plant. Do the roots try to contract to bring the bulb lower in the vase? Why or why not?

JOURNAL or BOTANY WORKBOOK

Do a section drawing of a bulb, preferably from one you've cut open so that you can examine it as you draw. See if you can find the stem that will hold the flower (called the peduncle). Sometimes you can even see an immature flower! Label. If you don't have a bulb yourself, watch Eileen's daffodil bulb dissection video.

Title a page in your joural "Not Roots" and collect names, pictures, and illustrations of plant parts that may look like roots but aren't. Label with the correct plant part. See if you can find examples of both stems and leaves. (hint: scale leaves)

MORE STEMS

Stems are often the trickiest plant parts to identify because they are so diverse. The basal plate stem in a daffodil bulb is just one example. In this lesson, we'll look at more unusual stems: rhizomes.

www.knightkingpress.com/botany

More Stems

Eileen Chevalier's TYPEWRITER Aalphabet BOTANY

NOTES

READ

Ultimate Visual Dictionary, DK (154-155)

DEFINE

rhizome - specialized stem that grows at or below ground, sending new roots down and new leaves up.

Why are rhizomes classified as stems?

QUESTION & INVESTIGATE

Dividing plants that grow on rhizomes is a fun way to share plants with friends. If you propagate an iris and share it, is the iris that your friend has the same as yours or slightly different? Why? Would differences happen because the DNA is different or because the plant is adapting to a new environment? What factors change how the iris grows?

Rhizomes are a special stem type. Examine how the rhizome branches. How is this similar to the stems that you can see above ground?

BONUS! Did you know that the Christmas Cactus doesn't have leaves? Those are actually stems. How can a plant survive without leaves? How do we know those are stems and not leaves? What does the flower attach to? What gives the plant structure?

JOURNAL or BOTANY WORKBOOK

Try to find fresh, soft-skinned ginger at your local market or grow your own. Make predictions in your journal about what will happen when you put a piece of fresh ginger on a dry counter and let it sit. Record how it changes each day. What differences can you observe in appearance, texture, weight, and smell? If you have a large enough piece, you can cut some off every few days and see how the taste changes as well. What is happening? Are the carbohydrates changing? Why does the texture change?

Draw or copy an illustration of rhizomes. Draw a line to show where the top of the soil is. Iris rhizomes are partially above ground, but ginger rhizomes are fully below.

MORE LEAVES

What makes a leaf a leaf? Why do so many people think the bract of the peace lily is a petal? In today's lessons, we'll look at the difference between leaves and try to figure out how to identify them.

www.knightkingpress.com/botany

More Leaves

Eileen Chevalier's TYPEWRITER Aalphabet BOTANY

NOTES

READ

Flora: Inside the Secret World of Plants, DK (108-181)

DEFINE

bract - modified leaf at the base of a flower or flower cluster that protects the buds and/or attracts pollinators

How is a leaf different from a petal? How can you tell the difference between petals and bracts?

QUESTION & INVESTIGATE

Reread the definition for leaf: flat structures extending from the stem that are responsible for photosynthesis. What modification happens to make a bract distinct from a common leaf? Why are bracts still considered a type of leaf?

Three types of bracts are: spathy, petaloid, and epicalyx. Examine a peace lily, poinsettia, and hibiscus. Can you identify the bract type? (*Flora*, 178)
Spathe is a large bract around a cluster of tiny flowers, often mistaken for a single petal.
Petaloid means petal-like. Look closer to find small flowers above the petal-like bracts.
Epi- has many meanings, including near, on, before, or after. The **calyx** is the group of sepals, located just under the petals. So, where could you find the epicalyx bracts?

In most plants, leaves are where most of the photosynthesis takes place. This is why they are green. What happens in the fall when leaves turn colors and fall from plants? Read *Flora* (p.154-155). Why don't leaves fall off when they are green? Do leaves "turn" colors? Where is the yellow color from?

JOURNAL or BOTANY WORKBOOK

Find three examples of bracts that have different jobs. Record them in your journal and explain what their main job is and if the bracts are mistaken for another flower part.

Although leaves look flat to us, they have many layers of cells that perform important jobs. Looks at the diagrams in *Flora*, pages 112-113, and copy the diagram. Don't worry if you don't know what everything is. Use this drawing in the Photosynthesis and Fluids lessons.

PETALS, SEPALS, AND TEPALS

Petals, sepals, and tepals are the showy or protective parts of a flower. They don't create seeds, but are an important part of the process. In this lesson, we'll learn the differences between petals, sepals, and tepals.

www.knightkingpress.com/botany

Petals, Sepals, & Tepals

Eileen Chevalier's TYPEWRITER Aalphabet BOTANY

NOTES

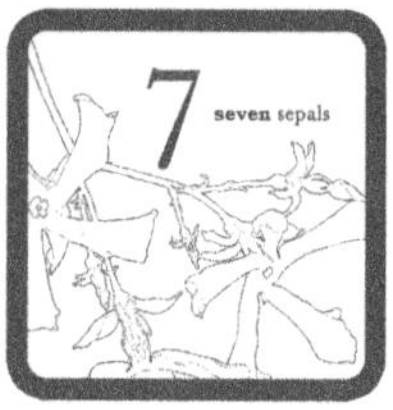

READ

Flora: Inside the Secret World of Plants, DK (184-185)

DEFINE

petal - showy flower part that attracts pollinators
corolla - all the petals of a single flower
   Do petal designs attract specific pollinators?
sepal - petal-like or leaf-like part that surrounds and protects the petals as they develop
calyx - all the sepals of a single flower
   Which falls off first: petals or sepals? Why?
tepals - petals and sepals that look and feel alike
   Which clade/s of flowers usually have tepals?
perianth - all petals (corolla) and sepals (calyx) of a flower
   What do "peri" and "anth" mean?

QUESTION & INVESTIGATE

Many orchid lovers refer to the inner tepals as petals and the outer tepals as sepals. Examine an orchid for yourself. Which term do you think is more accurate? Why? Is it ok that we use different terms or should we have one system that everyone agrees on and uses? Is it possible for everyone to agree on terms?

JOURNAL or BOTANY WORKBOOK

Examine the petals and sepals of several eudicots. Write down your observations about the size, shape, and texture of each and compare. How similar are petals and sepals? Draw a eudicot and label the petals, corolla, sepals, calyx, and perianth.

Examine the tepals of several monocots. Write down your observations about the size, shape, and texture of the inner and outer tepals and compare. (Inner tepals would be the petals; outer tepals would be sepals.) Draw and label the tepals and perianth.

Corolla Shapes

Eileen Chevalier's
TYPEWRITER
Aalphabet
BOTANY

COROLLA SHAPES

Botanists love to name things, and the shapes of a flower's petals (corolla) are especially fun. In this lesson we'll look at different ways to name the shapes of flowers.

www.knightkingpress.com/botany

NOTES

READ

Flora: Inside the Secret World of Plants, DK (188-189)

DEFINE

symmetrical - two halves of a bisected object having the same parts placed in the same places on both sides
radial symmetry - identical wedges meeting at a central point; cuts through the center result symmetrical halves
bilateral symmetry - only one set of halves is symmetrical

Do humans have radial or bilateral symmetry?
Are there parts of your body that have radial symmetry?

QUESTION & INVESTIGATE

Collect a variety of flowers (both monocots and eudicots) for the following activities:

Find a monocot flower with tepals instead of petals/sepals. What type of symmetry does it have? Next, take off the outer tepals (sepals). Does the type of symmetry change or stay the same? Why? Do you need the entire perianth to determine a flower's symmetry? Do the same with a eudicot.

Using the pictures in *Flora* (p. 188-189), determine the symmetry and shape of your flowers. Is there one shape that is the most common? (Make sure you look at the corolla and not the inflorescence! Sunflowers have a capitulum inflorescence with many tiny florets. Pull off and examine a single one.)

JOURNAL or BOTANY WORKBOOK

Make drawings of the flower shapes you discovered in this lesson (include some from *Flora*, too), then visit a park, nursery, or botanical garden to see if you can find more examples. If you see an unknown type, draw it so you can look up the shape name later.

Go through the Flashcards and see if you can identify the flower shapes. Hint: if the half of the flower that Eileen drew has a unique shape (ie. calcarate), the symmetry will be bilateral. Label some of the coloring sheets with corolla shape as well.

INFLORESCENCE

Humans have been creating flower arrangements for a long time, but flowers have been doing it themselves for far longer! In this lesson, we'll look at how angiospermae arrange flowers on their stems.

www.knightkingpress.com/botany

Inflorescence

Eileen Chevalier's TYPEWRITER Aalphabet BOTANY

NOTES

READ

Flora: Inside the Secret World of Plants, DK (207, 214-221)
Kk and Ww coloring sheets

DEFINE

inflorescence - how flowers are arranged on their stem

Can you discover the inflorescence without seeing the entire plant? Is it possible for a plant to have more than one type of inflorescence?

QUESTION & INVESTIGATE

Collect and/or observe a variety of flowers on their stems and identify which inflorescence type each has based on the pictures in *Flora* (pages 216-217). This is a great activity for a park or garden.

Read the Ee and Ss cards. What are the tiny flowers on a capitulum inflorescence called? Do all of them produce seeds? Why? If a floret can't produce a seed, is it really a flower? What makes a flower a flower? Are florets different because of their inflorescence?

Think about the order in which the flowers bloom along an inflorescence. Why don't they all bloom at the same time? Why might a staggered bloom time be beneficial for the plant? How would staggered bloom time affect plants with separate male and female flowers? (*Flora*, 207)

JOURNAL or BOTANY WORKBOOK

Use the Kk and Ww coloring sheets to compare bloom progression. Both vines have almost identical flower shapes, but is the inflorescence the same? Turn the Kk upside down. Label the shape and inflorescence shapes and include them in your journal.

Draw several inflorescence shapes. You can use bold lines to show the general shape or add details from the flower itself. Write the names of other flowers that have the same inflorescence next to your drawings.

REPRODUCTION

All living things must reproduce in order for their species to survive. In this lesson, we'll compare the two main ways angiospermae create new flowering plants: vegetative and sexual reproduction.

www.knightkingpress.com/botany

Reproduction

Eileen Chevalier's TYPEWRITER Aalphabet BOTANY

NOTES

READ

Ultimate Visual Dictionary, DK (144-155)

DEFINE

reproduction - creation of new living organisms using cells and genetic material from one or two existing organisms

What would happen if flowers couldn't reproduce?

DNA - deoxyriboneucleic acid, a string of genetic information that determines the appearance and function of each living thing.

QUESTION & INVESTIGATE

Some plants are masters of reproducing sexually. Have you ever seen dandelions pop up all over a yard that only had a few plants the year before? Other plants are masters of reproducing vegetatively. Have you planted a hosta or daylily and watched each year as more and more plants emerge in the spring? What makes a plant good at reproducing in one way or another? Is it possible for the same plant to reproduce vegetatively in one environment and sexually in another? Why?

Do all daughter plants from a single parent plant contain the same DNA? Many seeds contain genetic material from two different plants, but some plants are self-fertilizing and seeds are created from a single parent plant. In vegetative production, daughter plants have the same DNA as their single parent. Is that true of daughter plants created sexually (seeds) from a single part plant? Why? Is this always true?

JOURNAL or BOTANY WORKBOOK

Make a chart of the two ways angiospermae reproduce: vegetative and sexual. Compare the following: How many parent plants are needed? Will the new plant have unique DNA or the same DNA as the parent/s? What conditions are necessary for each type of reproduction to occur? Leave space to add more information from the lessons on vegetative and sexual reproduction.

VEGETATIVE REPRODUCTION

Angiospermae produce stunning flowers that make seeds, but most don't need them to reproduce. In this lesson, we'll look at how vegetative propagation allows plants to create replicas of themselves.

www.knightkingpress.com/botany

Vegetative Reproduction

Eileen Chevalier's TYPEWRITER Aalphabet BOTANY

NOTES

READ

Ultimate Visual Dictionary, DK (pages 154-155)
Flora: Inside the Secret World of Plants, DK (86-87; 104-105, 174-175)

DEFINE

vegetative reproduction - generation of new plants from the vegetative material (roots, stems, leaves) of existing plants

Can all vegetative parts of a plant create new plants?

propagation - increasing the number of a species through reproduction

How many ways can angiospermae propagate?

QUESTION & INVESTIGATE

Look at pages 174-175 in *Flora* to see a plant that reproduces using plantlets. What if the plantlets didn't grow adventitious roots before separating from the mother plant? If they hadn't grown roots yet, could they fall off or would they remain attached?

Look at the Not Roots and Rhizomes lessons and read pages 154-155 of the *Ultimate Visual Dictionary* to learn about more vegetative propagation methods. Discuss similarities and differences among the many ways plants reproduce vegetatively.

Propagate succulents from leaves. Try several leaves of several different genera (the plural of genus) if possible. Which take root? When do they produce new growth?

All but a few examples of vegetative reproduction in the lessons are based on stems. Why is that?

JOURNAL or BOTANY WORKBOOK

Make a chart to sort the information you've learned in this lesson. Which plants reproduce with which propagation methods? Are there plants that propagate using multiple vegetative propagation methods ?

What are advantages and disadvantages of vegetative reproduction? What if there were no pollinators (houseplants)? What if a particular trait kept a species from surviving a natural phenomenon (flood, wind...)?

SEXUAL REPRODUCTION

Flowers–the identifying characteristic of angiospermae–are where sexual reproduction takes place. In this lesson, we'll look at how flowers create seeds and explore the differences between self-pollination and cross-pollination.

www.knightkingpress.com/botany

Sexual Reproduction

Eileen Chevalier's TYPEWRITER Aalphabet BOTANY

NOTES

READ

Gregor Mendel: The Friar Who Grew Peas, Cheryl Bardoe
Ultimate Visual Dictionary, DK (154-157)
Flora: Inside the Secret World of Plants, DK (202-213, 268-269)

DEFINE

sexual reproduction - generation of new plants from the sexual parts (stamen, pistil) of existing plants

What are flowers for?

QUESTION & INVESTIGATE

Read *The Friar Who Grew Peas*. What are the advantages of sexual reproduction when certain traits are desirable or undesirable?

Sexual reproduction can happen within the flowers of one plant (self-pollination) or between the flowers of two different plants (cross pollination). There are advantages and disadvantages to each. Compare plants that avoid self pollination (*Flora*, 206-212) and those that use it (*Flora*, 268-269).

Although we enjoy their beauty, flower shapes aren't just for looking pretty. The flower's goal is to transfer pollen from the anthers to the stigma at the tip of the pistil. Read about the process of two-stage pollination on pages 202-203 of *Flora*. How does flower shape determine how a flower is pollinated? Why is this important?

JOURNAL or BOTANY WORKBOOK

Seeds are formed in the ovaries of flowers. Many are difficult to see, but look at the Dd, Hh, Ii, Ll, and Uu cards. Can you find the seeds forming? How many seeds are in each? Draw a picture of an ovary and label the seeds made by sexual reproduction.

Read and copy the "Process of Fertilization" diagram* in the *Visual Dictionary* (page 147). What would happen if the generative nucleus in the pollen grain didn't divide?

* The third picture states: "1st male gamete fuses with the ovum to form an embryo." Technically, it is a zygote until the first cell division when it becomes an embryo. See explanation on page 146.

PISTIL & STAMEN

The two sexual parts of the flower are the pistil (female) and stamen (male). In this lesson, we'll look at the structure of stamens and pistils as well as their relationship.

www.knightkingpress.com/botany

Pistil & Stamen

Eileen Chevalier's TYPEWRITER Aalphabet BOTANY

NOTES

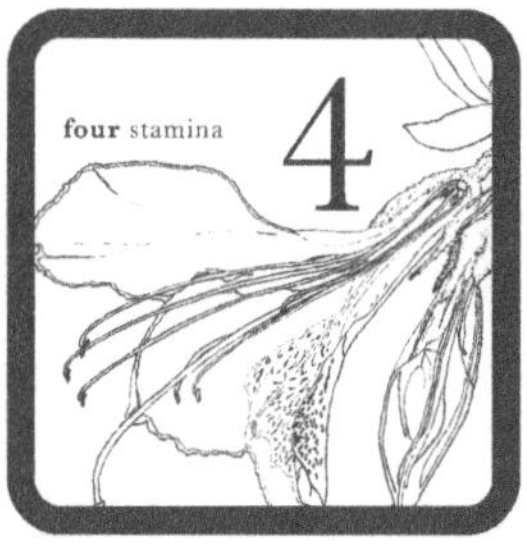

READ

Ultimate Visual Dictionary, DK (pages 144-145)
Flora: Inside the Secret World of Plants, DK (page 198)

DEFINE

stamen - male part of the flower (filament, anther)
pistil - female part of the flower (ovary, style, stigma)
Which part is in the very center of the flower? Why?

QUESTION & INVESTIGATE

Read the 3 and 4 Flashcards to learn the parts of the pistil and stamen. Can you identify the separate parts on the Flashcards? Open up some fresh flowers as well. What does the stigma feel like? Is there still pollen on the anther? Is it easier to bend/break the pistil or the stamen? Why?

Look through the Flashcards and compare the number of stamen and the number of pistils. (Each pistil has a single ovary, but can have more than one carpel. For instance, the iris has one compound pistil with three fused carpels. Roses have one very compound pistil - look at how many styles come up from the one fused ovary.)

Compare the ovaries of the Hh, Ll, and Qq cards. Where are the stamen located relative to the ovary? Which flower is hypogynous (hypo=under; gynous=ovary)? Which is Perigynous (peri=around)? Which is Epigynous (epi=above)?

JOURNAL or BOTANY WORKBOOK

Look a the magnified pollen grains on page 198 of *Flora* and pages 144-145 of the *Ultimate Visual Dictionary*. Draw a few of your favorites in your journal. Did you know that monocots and magnoliidae have 1 pore on their pollen grains and eudicots have 3?

Use the *Flashcards* to see if you can identify the pistil and stamens and their parts on each card. Then print out the Aa through Dd coloring pages and label the stamen and pistil parts. You might want to color code the pistil and stamens. Include these in your journal.

SEEDS & FRUIT

All angiospermae produce covered (angio) seeds (sperm) in their flowers. In this lesson, we'll look at how and where seeds are formed and compare succulent and dry fruits.

www.knightkingpress.com/botany

Seeds & Fruit

Eileen Chevalier's TYPEWRITER Aalphabet BOTANY

NOTES

READ

Ultimate Visual Dictionary, DK (pages 148-151)
Flora: Inside the Secret World of Plants, DK (pages 184-185, 300-301)

DEFINE

seed - unit that contains a plant embryo and is protected by a seed coat
fruit - the covering of an angiospermae seed

How are the ovary, seed, and fruit related?

QUESTION & INVESTIGATE

There are many ways to classify fruits. Simple fruits develop from a single ovary but may have many seeds; compound fruits develop from many ovaries but each may have a single seed. Cut open fruits and determine which are simple and which are complex.

Some fruits are called false fruits or pseudocarps. Examples are apples, figs (*Visual Dictionary*, 148), and rosehips. Often, fruits can develop from the receptacle of a flower (*Flora*, 184). How might the receptacle become the fleshy part of fruits? Are hypogynous, perigynous, or epigynous (see Pistil & Stamen video) flowers more likely to be false fruits? Why?

JOURNAL or BOTANY WORKBOOK

Look at pages 300-301 in *Flora*. Beside the name of each type, there is a simple line drawing of the fruit structure. Which are the most surprising or intriguing to you? What, technically, is corn? Draw a few structures, label with fruit type and an example.

The pericarp (fruit wall) can be succulent (peach) or dry (strawberry and pea). Some fruits are easy to identify. Peas are contained in a pea pod pericarp. Other times it is quite difficult. Strawberries aren't berries at all. In fact, they are dry fruits! How can this be? The key is the term pericarp. Remember that the ovary is part of the carpel (pistil). If the ovary forms into a fleshy covering (peaches), it is classified as succulent. If not, the ovary forms a dry covering. Pick some tricky fruits and record them in your journal.

LIFE CYCLE

Angiospermae begin life in a tiny seed and mature into a plant that can produce seeds of its own before dying. In this lesson, we'll look at the big picture of the life cycle of an angiosperm.

www.knightkingpress.com/botany

Life Cycle

Eileen Chevalier's TYPEWRITER Aalphabet BOTANY

NOTES

READ

A Seed is Sleepy, Diana Hutts Aston & Sylvia Long
Illustrated Dictionary of Biology, Usborne (pages 8-9)

DEFINE

life cycle - the developmental changes of an organism throughout its life: germination, growth, reproduction (seeds), and death

How are the life cycles of plants and humans similar?

QUESTION & INVESTIGATE

Read *A Seed is Sleepy* and compare the various life cycles in "A seed is generous." Why does it take some seeds so long to flower? How is the life cycle related to why pine nuts are so much more expensive than beans? Compare how long it takes seeds to germinate by reading "Some seeds are ancient" and looking at the picture of sprouting grass seeds in the lesson video. What other differences in life cycle have you discovered?

When reading about the life cycle of angiospermae, the cycle begins with a seed. What does this mean for the plants that are produced vegetatively? Are they still part of the parent plant even if the parent dies? Did they begin their own life cycle? When? Do they have a life cycle that looks different from plants that came from seeds?

JOURNAL or BOTANY WORKBOOK

The best way to examine the life cycle of a plant is to see it in action by growing plants yourself. When you buy seeds, look for a short "days to maturity" number. Fruits that are usually called vegetables often have short life cycles. Try cucumber, okra, or tomato.

Keep records of your life cycle growing experiment. What day did you plant the seeds? How many? How many seeds germinated? How many seeds developed true leaves? How many blooms are on each stem? What pollinators visit your flowers? How long did it take for the seeds to mature? Did the plant die after it stopped blooming? How long did each step take? If you grow an edible flower, taste it and record your observations!

POLLINATORS

While some angiospermae rely on wind or the anatomy of their flowers for pollination, many rely on help from the animal kingdom: bees, wasps, moths, butterflies, and more. In this lesson, we'll discover the vast array of pollinators.
www.knightkingpress.com/botany

Pollinators

Eileen Chevalier's TYPEWRITER Aalphabet BOTANY

NOTES

READ

Flowers are Calling, Rita Gray and Kenard Pak
Flora: Inside the Secret World of Plants, DK (212, 227-267)

DEFINE

pollen - tiny grains attached to the anther of a stamen; contain DNA that can combine with an ovum's DNA to create new plants
pollination - the transfer of pollen from the anther to the stigma of a pistil
pollinator - animal that aids in pollination
Can humans be pollinators?

QUESTION & INVESTIGATE

It takes work to attract a pollinator. How do plants keep them there? *Flora* give some interesting examples (pages 212 and 258). Without resorting to extremes, most flowers offer something in exchange for pollination. What is it? Where is it?

The Ss and Ee flashcards show examples of the capitulum inflorescence. (capit=head; inflorescence=how flowers are arranged on a stem) In this type of arrangement, the florets bloom progressively, sometimes over the course of weeks. (*Flora*, 214) How does this help the plant attract and keep pollinators? How does this prolonged blooming compare to other flower inflorescences? Do most flowers bloom all at once or continually over time? What are the advantages and disadvantages of each?

Read pages 234-5 of *Flora* to learn about "buzz pollination." How is this different from the way most flowers are pollinated? How is it similar to wind pollination? Could any other pollinator accomplish pollination for this plant?

JOURNAL or BOTANY WORKBOOK

Create a journal entry with some of your favortie pollinators and the plants they love.

Create a more detailed entry that illustrates the specific relationships between certain plants and their pollinators. Many examples can be found in *Flora* (pages 227-267).

NECTARIES

Pollinators aren't just helpful gardeners. They are in search of food for themselves and their colonies. Nectar is what draws most pollinators. In this lesson, we'll search out the secrets of the hidden nectaries.

www.knightkingpress.com/botany

Nectaries

Eileen Chevalier's TYPEWRITER Aalphabet BOTANY

NOTES

READ

Ultimate Visual Dictionary, DK (pages 144-145)
Flora: Inside the Secret World of Plants, DK (pages 250-253, 264-265))

DEFINE

nectar - sugary liquid produced by flowers to attract pollinators
nectaries - special plant parts that store nectar
nectar guides - markings on flowers that lead to the nectaries

Do all flowers make nectar?
Is all nectar stored in nectaries?

QUESTION & INVESTIGATE

How and where is nectar made? Why would a plant spend so much of its resources creating food for animals instead of using that food for itself? What is the flower trying to accomplish? How is this relationship a win-win? Could it also be called survive-survive?

Nectar guides can be beautiful for us to look at, but did you know that there are many nectar guides we can't see? What can see them? Look at examples of UV guides in the *Visual Dictionary* (144-145) and *Flora* (250-253, 264-265). Why might flowers use colors that we (and maybe other animals) can't see? Why might it use both?

JOURNAL or BOTANY WORKBOOK

Look through the *Flashcards* and identify any nectaries or nectar guides. Why can't you find them in every picture? Print coloring pages and label the nectaries and guides. Include the flower's pollinator/s if you know. (nectaries: Nn/1, Tt/9; guides: Aa, Hh, Ii, Oo,* Tt, Uu, and Xx) *Orchids are very deceptive! They may have nectar guides but no nectar.

Gather living flowers with visible nectaries and draw the flower. Label nectary, flower name, pollinator/s if known, and observations about nectar guides, color, and scent.

EDIBLE ANGIOSPERMAE

Enough theory. What do edible flowers taste like?! Most of the fruits and vegetables you already eat are from flowering plants: cucumber, pineapple, coffee, cilantro... In this lesson, try tasting some flowers as well!

www.knightkingpress.com/botany

Edible Angiospermae

Eileen Chevalier's TYPEWRITER Aalphabet BOTANY

NOTES

READ

100 Edible & Healing Flowers, Margaret Roberts
The Herbal Handbook for Homesteaders, Abby Artemisia

DEFINE

edible - safe for humans to eat

Can a plant be both edible and inedible?

QUESTION & INVESTIGATE

Did you know that cashews are toxic unless roasted and steamed? Many common foods are from plants with toxic parts or forms: tomatoes (vines and leaves), potatoes (when green), kidney beans (if not properly boiled), and asparagus (fruit). It's very important to know when and how to eat plants, including the flowers. How can you discover how to select and prepare angiospermae?

There are many ways to consume angiospermae: raw, cooked, as tea, and even candied. Look through the books above or online and come up with some recipes for a flower eating party. Which are your favorite? Can you grow your own flowers to eat so that they're most nutritious?

JOURNAL or BOTANY WORKBOOK

Try eating some flowers and record your impressions! It's even more fun in a group - you can take polls about taste, vote for your favorite flowers and varieties, and just watch the reactions of others to something new. Try eating several different types of nasturtiums and compare. Which tea is best hot or iced: echinacea, hibiscus, or lavender? Record your observations and rankings in your journal or workbook.

PROTECTION

All living things strive to stay alive. Many animals, including humans, do this by eating plants, but the plants need to stay alive, too. In this lesson, we'll discuss visible (physical) and invisible (chemical) methods of protection.
www.knightkingpress.com/botany

Protection

Eileen Chevalier's
TYPEWRITER
Aalphabet
BOTANY

NOTES

READ

Flora: Inside the Secret World of Plants, DK (78-79, 98, 156-158)
Wicked Plants: The Weed that Killed Lincoln's Mother & Other Botanical Atrocities, Amy Stewart

DEFINE

protect - to keep safe from harm or injury
How do animals protect themselves from plants?

QUESTION & INVESTIGATE

The most obvious way plants defend themselves is with physical obstacles. Thorns are often the first things that come to mind. Can you think of plants with thorns? If you said roses, you are mistaken. What?! Yes! Read the back of the 2 Flashcard and *Flora* (78-79, 98, 158) to learn the difference between spines, thorns, and prickles. What do you think would be more damaging to a plant: removing thorns or removing prickles? Why?

Plants also have invisible ways to keep predators at bay. Many plants have at least some poisonous parts and many deter pests with their bitter taste. Read about the habanero chili (51-53) and manchineel tree (99-100), and others (95-97) in *Wicked Plants*.

Animals aren't the only threat to plants. Some plants (Ff, Ss) release substances into the ground to keep other plants from growing. This is called allelopathy. Why do some plants use this technique?

JOURNAL or BOTANY WORKBOOK

Document defense mechanisms in two groups: visible and invisible.
What about stinging leaves (*Flora*, 156)? Can we see them or not? Have you ever touched a leaf only to discover it was covered with tiny points?

Make up your own plant with defense mechanisms. What is it protecting itself from? How will it still ensure pollination. Draw, label, and name your plant!

POISON & MEDICINE

Have you ever had to take medicine and been warned not to take too much? Why? What would happen? In this lesson, we'll take a closer look at plants that can both heal and kill. It's all about delivery and dosage!

www.knightkingpress.com/botany

Poison & Medicine

Eileen Chevalier's TYPEWRITER Aalphabet BOTANY

NOTES

READ

Wicked Plants: The Weed that Killed Lincoln's Mother and Other Botanical Atrocities, Amy Stewart

DEFINE

poison - substance that causes physical harm or death
Can you be poisoned just by touching a plant?

QUESTION & INVESTIGATE

Read the Xx Flashcard. What is the scientific name for foxglove? How does this relate to the name of the chemicals found in foxglove that are used for medicine: digoxin and digitoxin? The Latin word for finger is digit and the foxglove got its scientific name because it looks like a thimble. Which came first: the scientific name for foxglove or the name of the chemicals used in medicine? How do you know?

Are all medicinal plants, even common ones–turmeric, tea tree, lavender–also capable of being poisonous? Are all plants capable of being poisonous? What would happen if you took too many vitamins at once?

In the Protection lesson, we learned that being poisonous is a protective trait. Many common plants are poisonous, but there are very few we currently use as medicine. What would be obstacles and benefits to discovering the potential medicinal qualities of poisonous plants? Is there a safe way to test new medicines?

JOURNAL or BOTANY WORKBOOK

Create a poster promoting plant safety. What are some ways that you can stay safe around plants? Think about hiking in the woods (poison ivy), houseplants that don't want to be touched (dumb cane), and even being careful around flower arrangements (from the Flashcards, especially Bb, Hh, Ll, and Xx).

Read some of the chapters in *Wicked Plants* and create a "Wanted" poster for one plant you find particularly strange or terrible. Identify the plant name, picuture, and description of what it does. Include historical uses if possible.

DECIDUOUS & EVERGREEN

The landscape in winter can look very different from the landscape in spring, especially in cold and temperate climates. Why? In this lesson, we'll look at the differences and similarities between deciduous and evergreen plants.
www.knightkingpress.com/botany

Deciduous & Evergreen

Eileen Chevalier's TYPEWRITER Aalphabet BOTANY

NOTES

READ

Flora: Inside the Secret World of Plants, DK (p. 110-111)
Grow Your Soil, Diane Messler (Chapters 2-3; page 110)

DEFINE

deciduous - plants that lose all their leaves annually
evergreen - plants that retain leaves throughout the year

Do the individual leaves of an evergreen remain green?
Can a plant function as both deciduous and evergreen depending on the climate?

QUESTION & INVESTIGATE

Is a leaf evergreen or is the plant evergreen? How can you find out? Look at the base of evergreen plants to see if there are any leaves there. Are they green or did they turn color before they dropped? Why did the leaves drop? What happens to a leaf that can't do its job of photosynthesizing? Why can't some leaves photosynthesize?

Can leaves nourish a plant after they fall? Find a plant with decaying leaves at its base and push away the leaves to see what's happening. How would the soil be different if the leaves weren't there? What does a plant need? How do decaying leaves help the plant obtain water and nutrients? Read Diane Messler's *Grow Your Soil!* to find out what's happening when leaves decay on the soil and why it's so important for plants.

JOURNAL or BOTANY WORKBOOK

Collect a variety of leaves from evergreen and deciduous plants. How are they similar or different? Examine their shape, size, sheen (how shiny), and thickness. Record your observations and try pressing the leaves into your journal.

Carefully observe a plant when it's putting out new leaves. Ideally, use a plant that you see often so that you can draw and record observations of the leaf development. Some plants will have many leaves in different stages of development, so you can examine the entire process at once.

ANNUAL & PERENNIAL

Sadly, all plants will eventually die. Why do some only live for a few months and others continue to grow for hundreds of years? In this lesson, we'll look at annuals, perennials, and plants that can be both!

www.knightkingpress.com/botany

Annual & Perennial

Eileen Chevalier's TYPEWRITER Aalphabet BOTANY

NOTES

READ

Grow Your Soil, Diane Messler (Chapters 2-3, p. 119-120)
Flora: Inside the Secret World of Plants, DK (p. 190-191)

DEFINE

annual - plant that completes its life cycle in one year
perennial - plant that lives beyond one year, often not mature enough to produce seeds for several years

Could a plant be both annual and perennial?

QUESTION & INVESTIGATE

Read your Yy & Zz cards. How are the words perennial and annual related? What words can you think of with the same Latin roots?

Ecosystems need a variety of annuals and perennials. Why? What would happen to animals that need to eat early in the year if there were only perennials that bloomed in the summer? How do staggered bloom time and life cycle length support the entire system?

Visit a garden center and find tags that give the zone (area) where plants are hardy (survive the winter). Plants that are hardy in warmer areas can be sold in colder areas to be grown as annuals or as houseplants. Can you find plants that aren't hardy in your area?

How does the type of compost affect the soil microbes? Do annuals or perennials prefer bacterially-dominated compost? Which prefer fungally-dominated? How can you ensure that plants get the type they need? How does this happen in nature?

JOURNAL or BOTANY WORKBOOK

Grow an annual angiosperm (nasturtium, sunflower, zinnia) from seed and record its life cycle. Collect leaves, flowers, and seeds to press into your journal.

Examine perennial stems to discover where last year's growth meets this year's. New growth is usually smooth and green. Compare the stems of annuals and perennials. Why is a bleeding heart stem like an annual? Record pictures and observations in your journal.

CELLS 1

From zygotes with just one cell to the trillions in humans or mature trees, cells work together to complete everything required for life. In this lesson, we'll look at how energy works within the atoms that create cells.

www.knightkingpress.com/botany

Cells 1

Eileen Chevalier's TYPEWRITER Aalphabet BOTANY

NOTES

READ

Ultimate Visual Dictionary, DK (pages 308-311)

DEFINE

cells - tiny building blocks that link together to build all living things
atoms - tiny balls of energy that are the building blocks of cells

How many atoms are in a cell? How may cells in a plant?

QUESTION & INVESTIGATE

Atoms connect to create molecules. Molecules connect to create the structures that form organelles, which work together as cells. But how do all the atoms stick together?

ATOMS NEED TO BE NEUTRAL. Energy is called "positive" (+) or "negative" (-).
Protons (+) are in the center (nucleus) of the atom and electrons (-) orbit the nucleus.

ATOMS SEEK FULL SHELLS. Atoms like having full shells even more than being neutral, so they often gain or lose electrons and thus become positively or negatively charged.

OPPOSTIES ATTRACT. Postively charged atoms (cations with missing electrons) cling to negatively charged ones (anions with extra electrons).
The way in which atoms cling to one another creates the form and function of molecules.

Do you need to know about atoms to understand cells? How does a cell do work, such as harness energy from the sun, form and destroy protein molecules, form and reform the strands of cytoskeleton, or monitor the molecules passing through its cell membrane?

How/where is the energy of the sun turned into the energy that cells use to do work?

JOURNAL or BOTANY WORKBOOK

Examine the diagrams on pages 308-9 of *Visual Dictionary*. Make your own drawing or copy some from the book. Take notes from the video lesson and record any questions you have. How does energy work? (Don't worry, scientists all over the world are still trying to answer this quesiton. How can your continuing investigations contribute?!)

CELLS 2

Cells are the building blocks of all living things and each cell has a particular job within an organism. In this lesson, we'll learn about organelles in general and a few special organelles in more detail.

www.knightkingpress.com/botany

Cells 2

Eileen Chevalier's TYPEWRITER Aalphabet BOTANY

NOTES

READ

Illustrated Dictionary of Biology, Usborne (10-12, 26)

DEFINE

organelle - interdependent (distinct yet dependent on one another) structures in a cell that perform specific tasks

How is a cell like a miniature city?

QUESTION & INVESTIGATE

Read your Ww card to see how organelles function like parts of a city. Sometimes it's easier to remember what organelles do if you think of them as people or places where you live. Read about more organelles in *Usborne's Illustrated Dictionary of Biology*. Can you think of what to call the other organelles?

Most organelles are enclosed in a phospholipid membrane. Why would they need this? What is the cytoplasm that the organelles are floating in like? Lipids are fats, like oil. What do they do in water? The lipid ends of the molecules cling together so the phospho ends are packed together on the outside and inside of the membrane. This forms a layer with enough strength, flexibility, and thickness that organelles can function without harm, add openings as needed, and alter their shape and size. Which organelle needs its membrane to protect the rest of the cell from destruction? Which has a double membrane?

JOURNAL or BOTANY WORKBOOK

Draw a diagram of a cell with examples of organelles. Which organelles are the most plentiful? Where are the chloroplasts (page 26, *Usborne's Illustrated Dictionary*)? Make sure to label your drawing with the names and function of the organelles.

Which is the most important organelle? How many other organelles depend on that one? Make a chart or drawing to illustrate how the organelles are interdependent. Which organelles need the nucleus to tell them what to do? Which organelles need the lysosome to get rid of the junk that's in their way? Which organelles need chloroplasts to produce the energy they need?

PHOTOSYNTHESIS

Animals eat plants, but what do plants eat? In other words, how do plants get energy? The sun! In this lesson, we'll learn about photosynthesis, how plants capture energy from the sun and store it in glucose molecules.

www.knightkingpress.com/botany

Photosynthesis

Eileen Chevalier's TYPEWRITER Aalphabet BOTANY

NOTES

READ

Ultimate Visual Dictionary, DK (pages 138-139)
Flora: Inside the Secret World of Plants, DK (pages 38-39, 102-103, 129)

DEFINE

photosynthesis - process of converting energy from the sun into food
Where does the engery humans use come from?

QUESTION & INVESTIGATE

Glucose (sugar) has three types of atoms: 6 carbons (C), 12 hydrogens (H), and 6 oxygens (O). You can write it is like this: $C_6H_{12}O_6$. Look at the photosynthesis diagram in the Visual Dictionary. You can write the process of photosynthesis like a math problem:

Energy (sunlight) + CO_2 (carbon dioxide) + H_2O (water) = $C_6H_{12}O_6$ (glucose) + O_2 (oxygen) + H_2O (water)

You need (6) CO_2 moleules and (12) H_2O moleules to make one glucose because of how the energy is captured. Where does the energy that went into the reaction go?
The opposite of capturing energy is respiration. The equation for respiration is the photosynthesis equation in reverse. How do human cells get energy to stay alive? What do we breathe out? How do plants get energy? Do plants ever "breathe out" CO_2?

Photosynthesis involves a lot of gasses (carbon dioxide and oxygen) flowing in and out of leaves. How do gas molecules get into the leaves? Look at the diagram. (*Ultimate Visual Dictionary*, 138; *Flora*, 129) What would happen if the leaves were covered in oil? Why?

JOURNAL or BOTANY WORKBOOK

Most photosynthesis takes place in leaves, but some plants photosynthesize in stems and even roots! What kinds of plants do you think they could be? What if a plant didn't have any leaves? Where would roots have to be to photosynthesize? Read about and document two special plants in *Flora* (38-39, 102-103).

Draw a diagram of how photosynthesis works. You can use pictures, words, lines, and numbers. See if you can use your diagram to explain photosynthesis to a friend.

TRANSPORTATION OF FLUIDS

The most common fluid in the world is water. No living thing can survivie without it. How do plants "drink" water and move it around? In this lesson, we'll look at how plants move and use water and other fluids.

www.knightkingpress.com/botany

Transportation of Fluids

Eileen Chevalier's TYPEWRITER Aalphabet BOTANY

NOTES

READ

Illustrated Dictionary of Biology, Usborne (14-21, 24-25)

DEFINE

osmosis - flow of water from a high concentration area to a low concentration area though a semi-permeable membrane

capillary action - movement of water through very narrow tubes, allowing water to flow in opposition to gravity

Do plants need both osmosis *and* capilary action?

QUESTION & INVESTIGATE

Fluids are moving through plants all the time. The two transport systems are xylem and phloem. Xylem uses capilary action, drawing water from the roots, through the stems, and to the leaves. Phloem moves fluid throughout the plant, even down to the roots. One system mainly moves minerals and the other mainly moves sugars. Which is which? Where do the sugars and minerals come from? Where do they need to go? Why can't there just be one system that flows from the roots to the leaves?

Transpiration is when water is pushed out of plants through tiny opening in their leaves called stomata. If water can exit this way, can it also be absorbed through the stomata? Why do some plants need a moist environment to stay healthy?

JOURNAL or BOTANY WORKBOOK

The cell wall is important for plant shape, but cells depend on vacuoles to avoid wilting. See what happens when the vacuole loses too much fluid (*Illustrated Dictionary*, 25). Learn about osmosis and create an illustration of how fluids move into and out of cells.

Create a diagram of an entire plant: root, stems, and leaves. Illustrate the movement of fluids through xylem and phloem, osmosis, and transpiration. Also include dangers to the transport system: too much or too little water at the roots, wind/sun, aphids stealing sap, covered stomata, etc.

IDENTIFICATION

In this lesson, we'll review the visible characteristics of plants that we learned about in this course. Then you can do some fieldwork! Go outside to find different plants and see what characteristics you can identify.

www.knightkingpress.com/botany

Identification

Eileen Chevalier's TYPEWRITER Aalphabet BOTANY

NOTES

READ

Typewriter Aalphabet Flascards Botany Journal
Field guide for identifying plants in your area
(If visiting a greenhouse, find a guide for the native habitats)

DEFINE

fieldwork - working in the field (real world) to gain experience and knowledge through observation and examination

Why is fieldwork important?

QUESTION & INVESTIGATE

You've learned so much about the physical characteristics, growth, and development of different types of plants. Read through your journal to refresh your memory and make a list of the major characteristics of different types of plants.

How can you identify an angiospermae?

What are the differences between monocots, eudicots, magnoliids, & basal angiosperms?

What clues do flowers, stems, and leaves give us?

What can we discover from the height of a plant?

JOURNAL or BOTANY WORKBOOK

Fieldwork! Grab your journal, writing utensils, loupe, camera, and a field guide for plants in your area. Plan to spend some time at a botanical garden, park, the woods, or even your own backyard. See how many plants you can identify. Record them as well as ones that you want to identify later. Pictures, drawings, and sample leaves can help. Happy identification!

Aa
Azalea

Bleeding Heart

Christmas Cactus

Cc

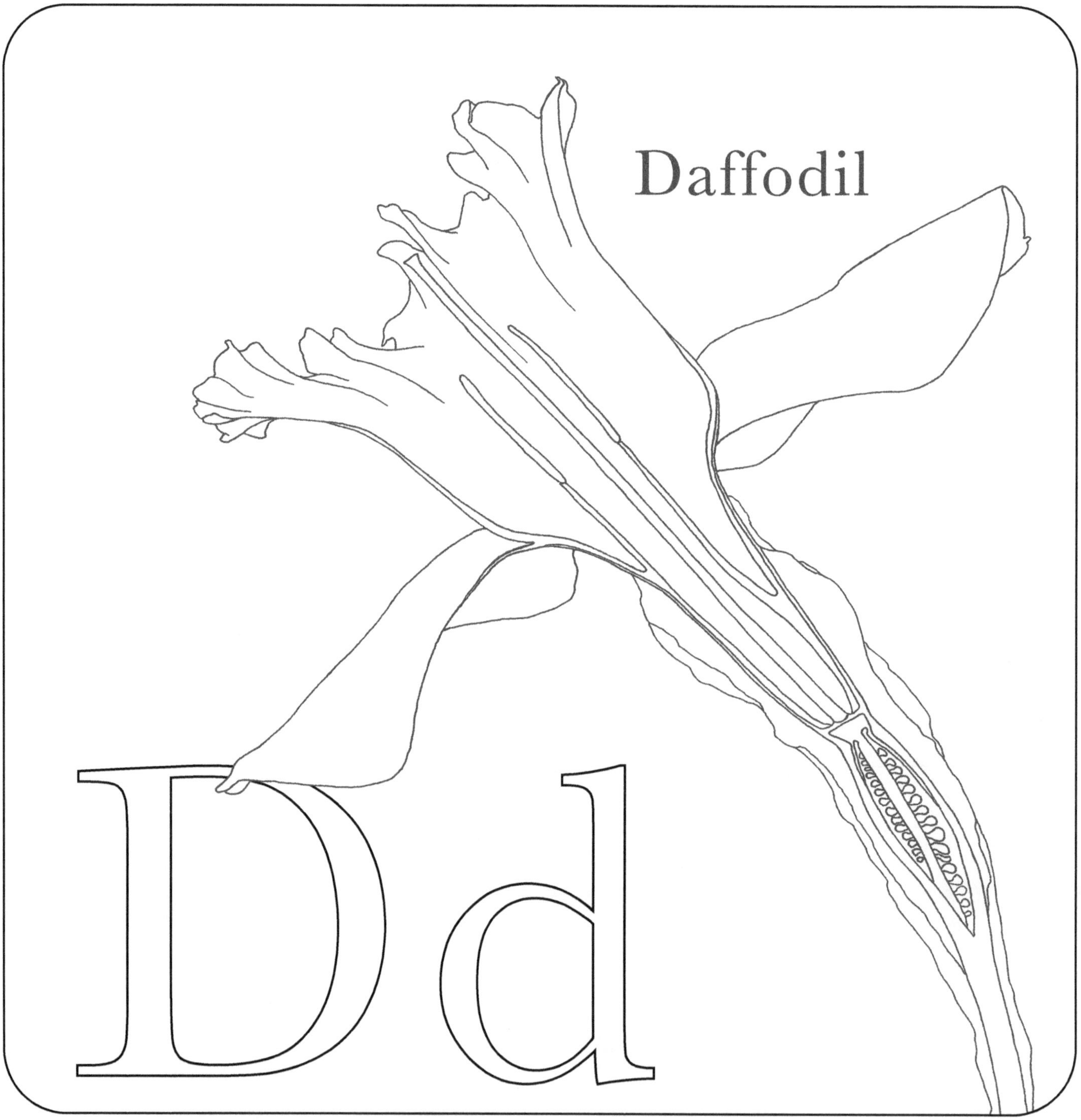
Daffodil
Dd

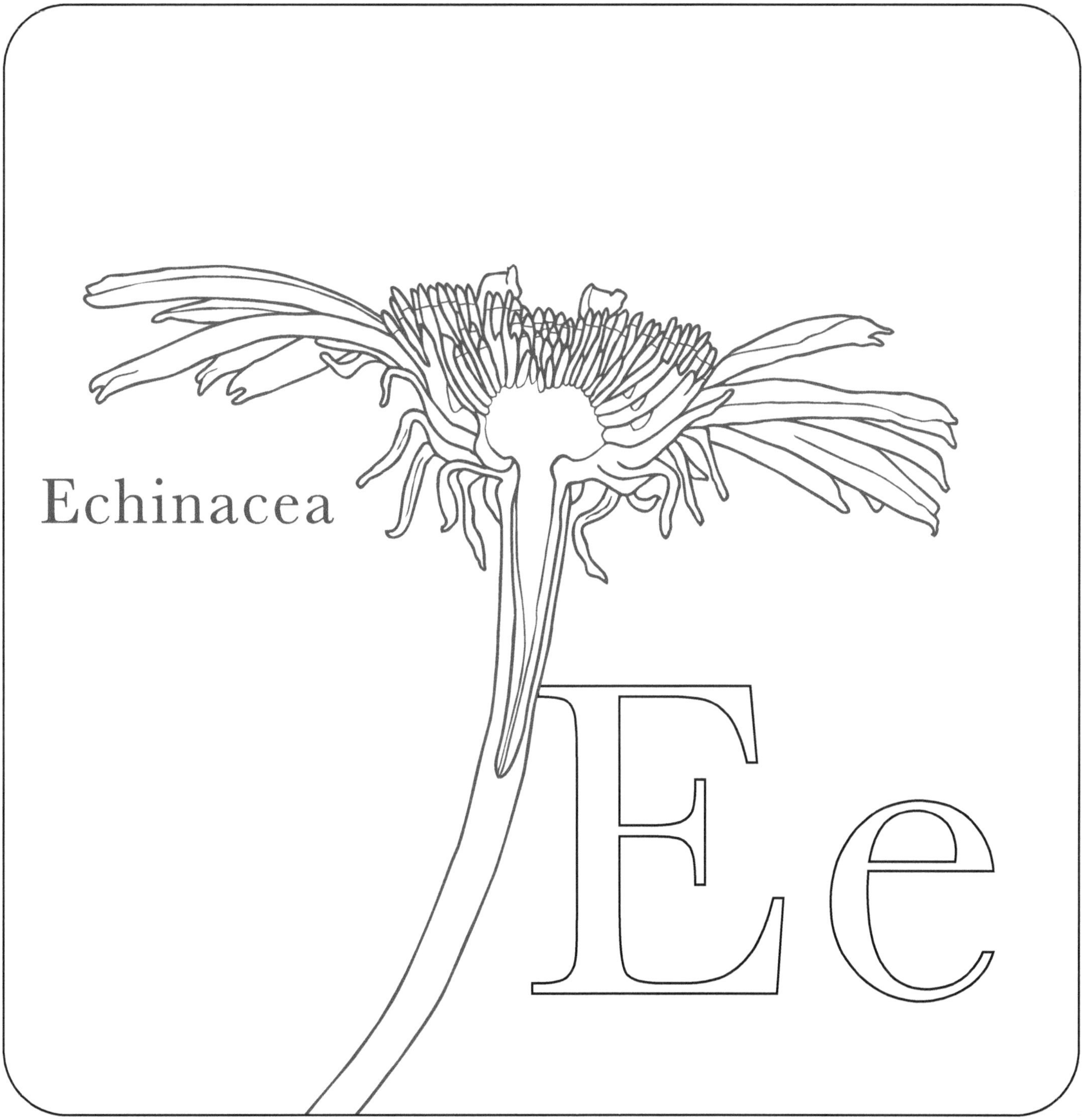
Echinacea
Ee

Ff
Forsythia

Gardenia
Gg

Hellebore

Hh

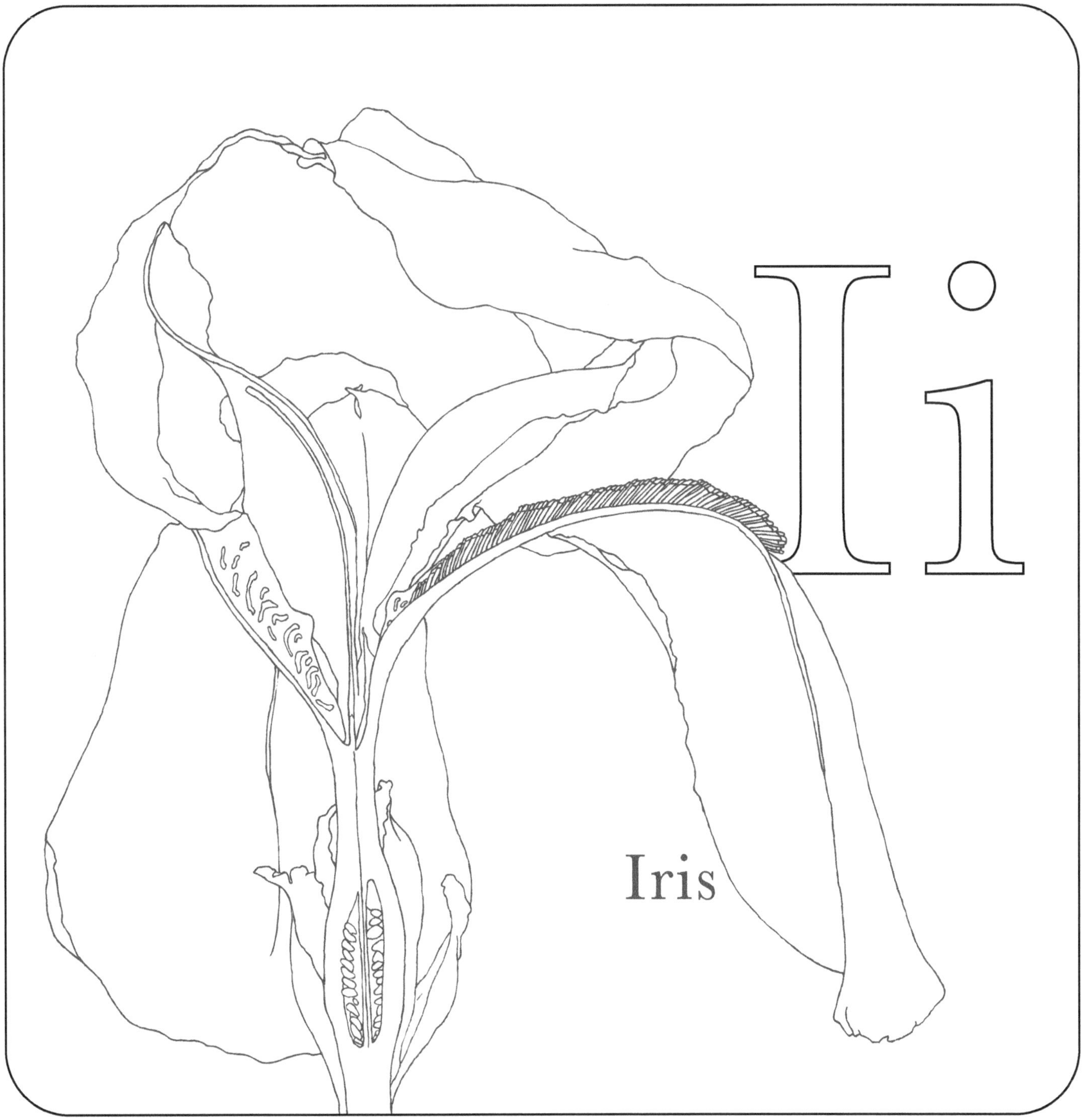
Ii
Iris

Jj
Jasmine

Kudzu
Kk

Lily of the Valley

Ll

Magnolia
Mm

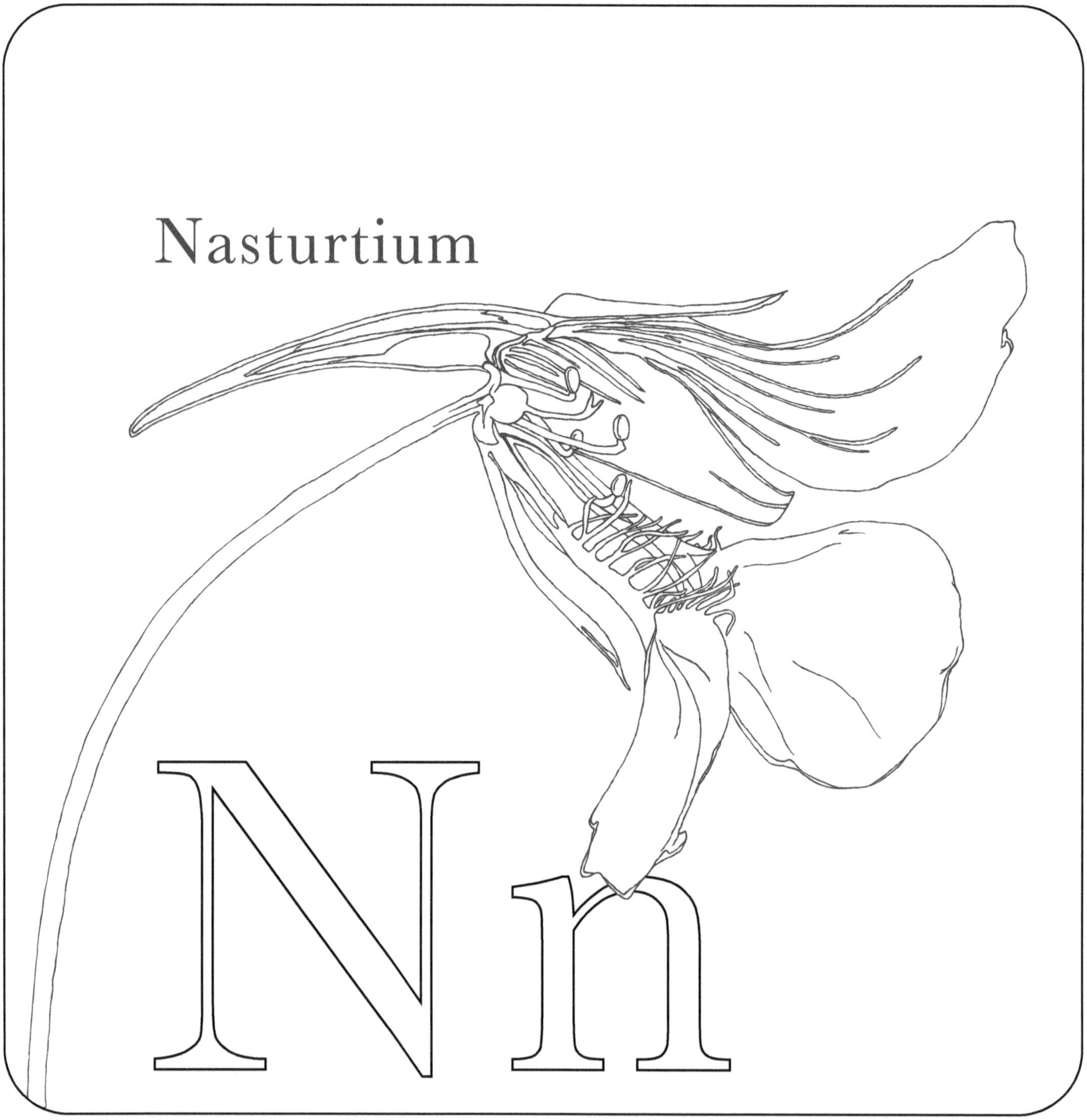
Nasturtium
Nn

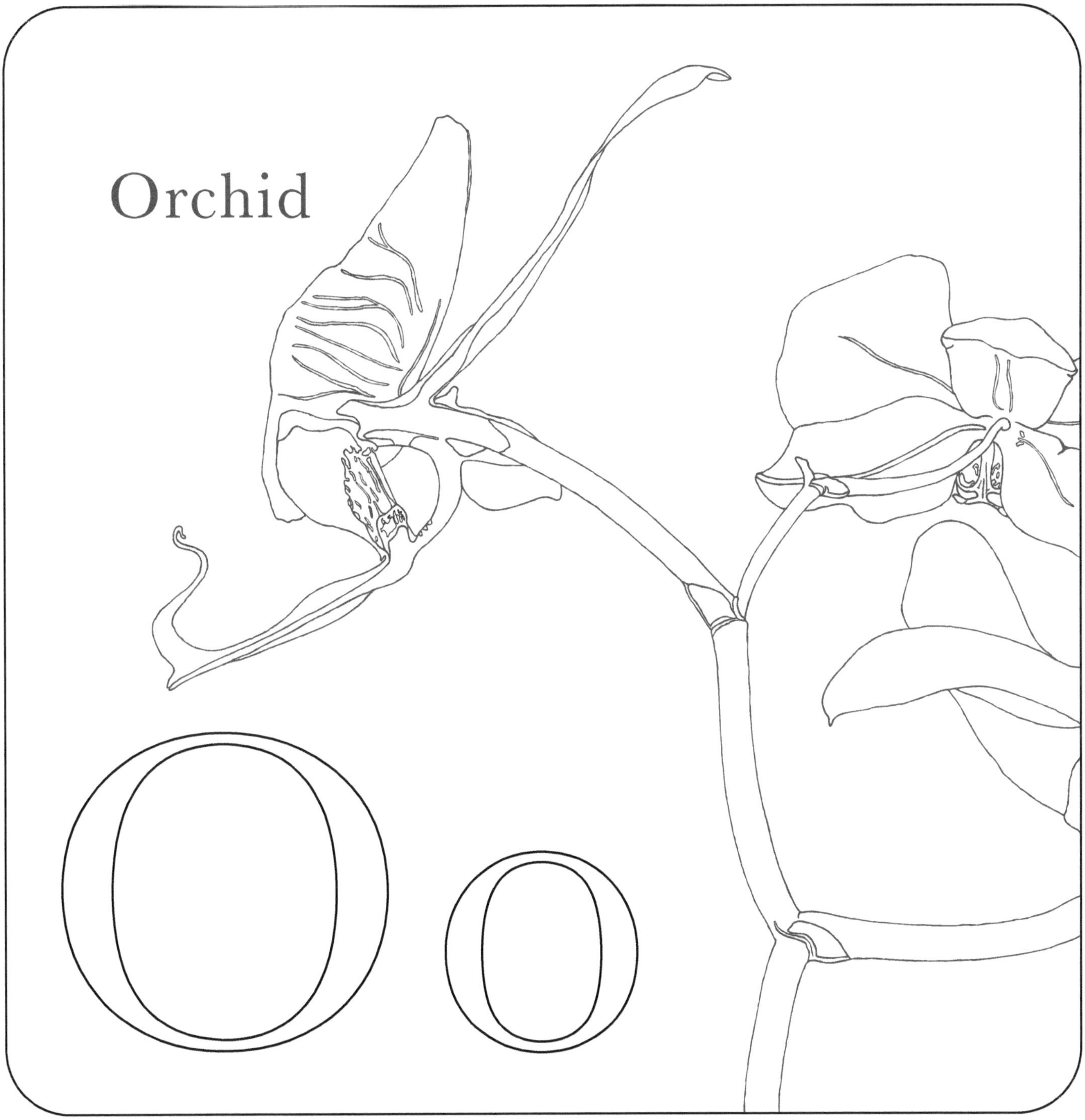

Peach
Pp

Qq

Quince

Rose
Rr

Ss

Sunflower

Tt

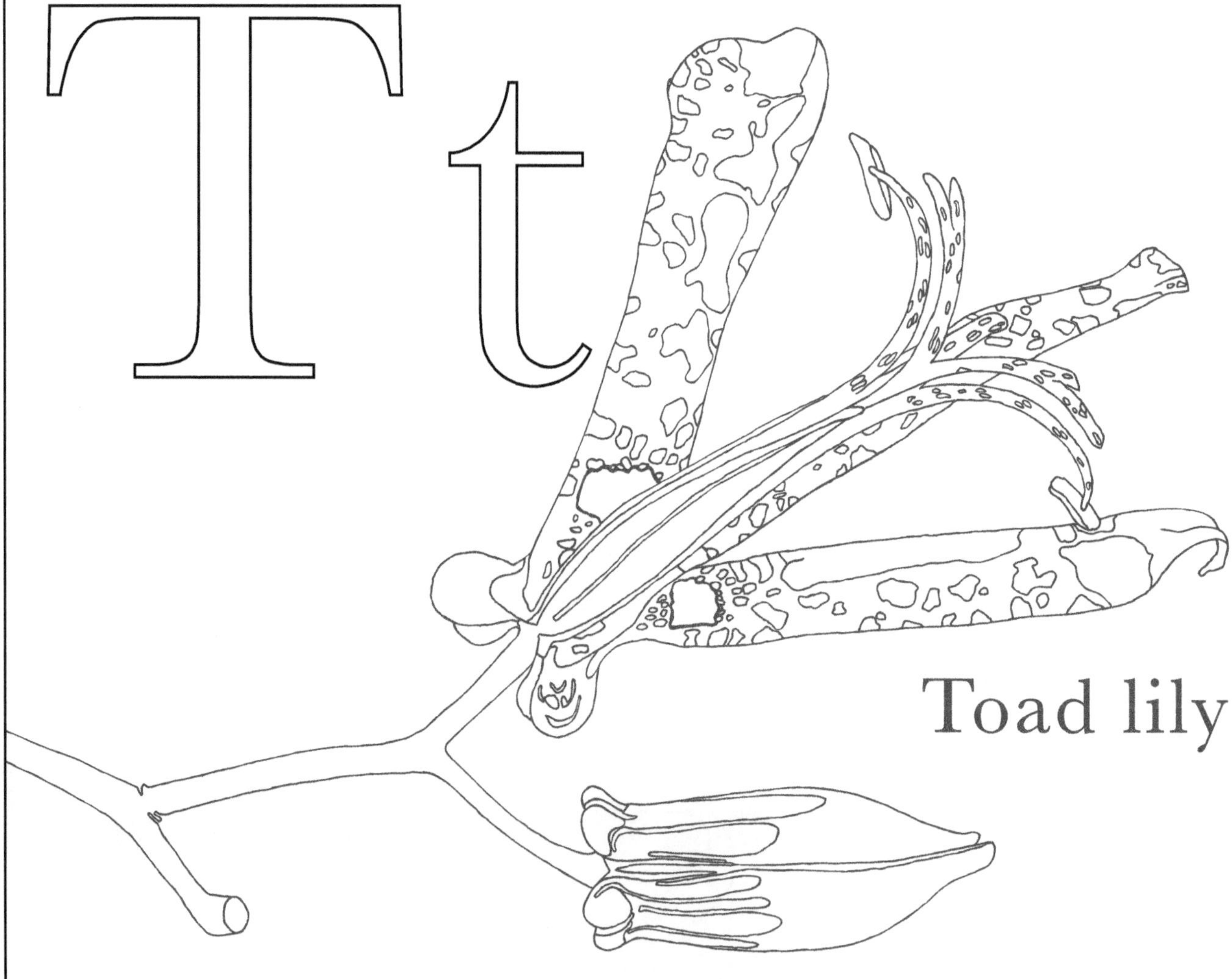

Toad lily

Uu

Ulster Mary

V v

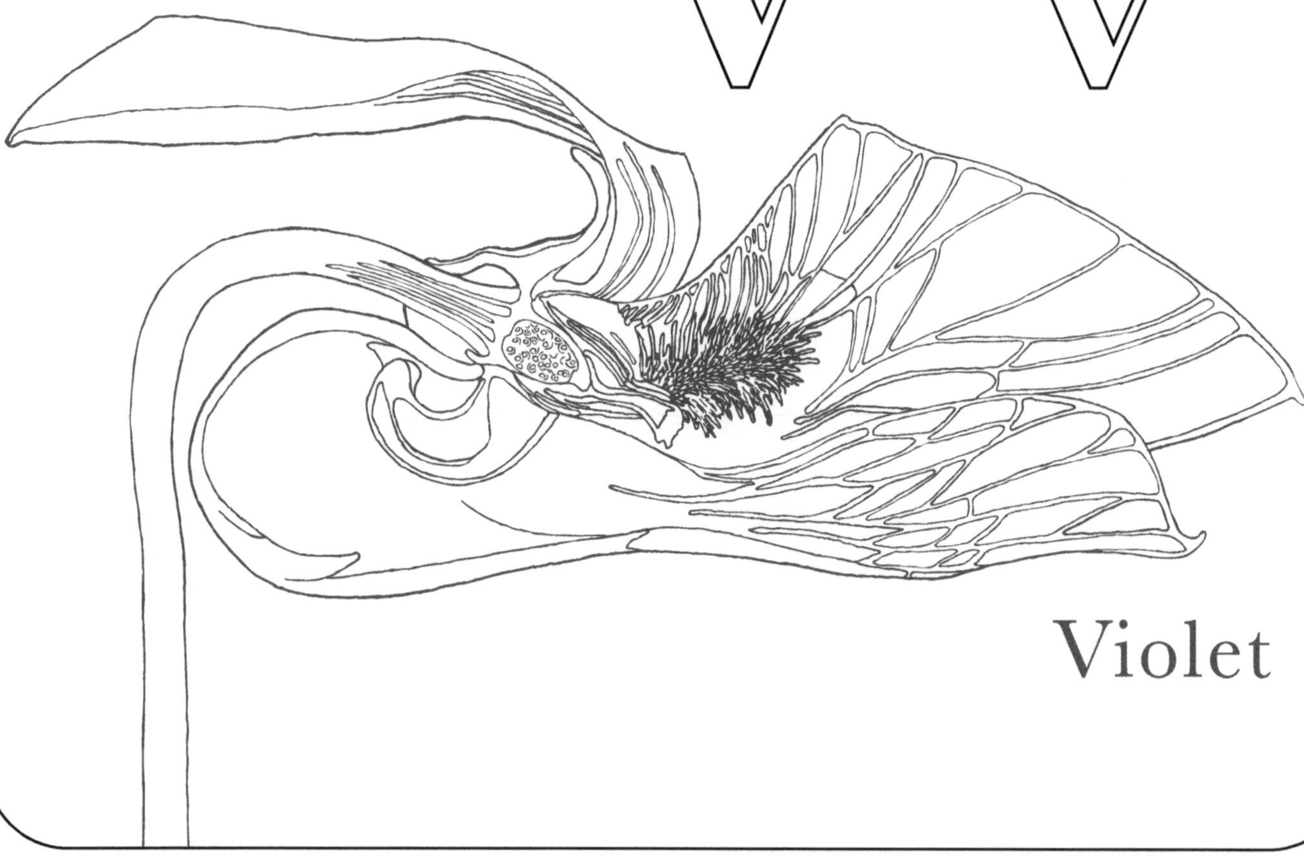

Violet

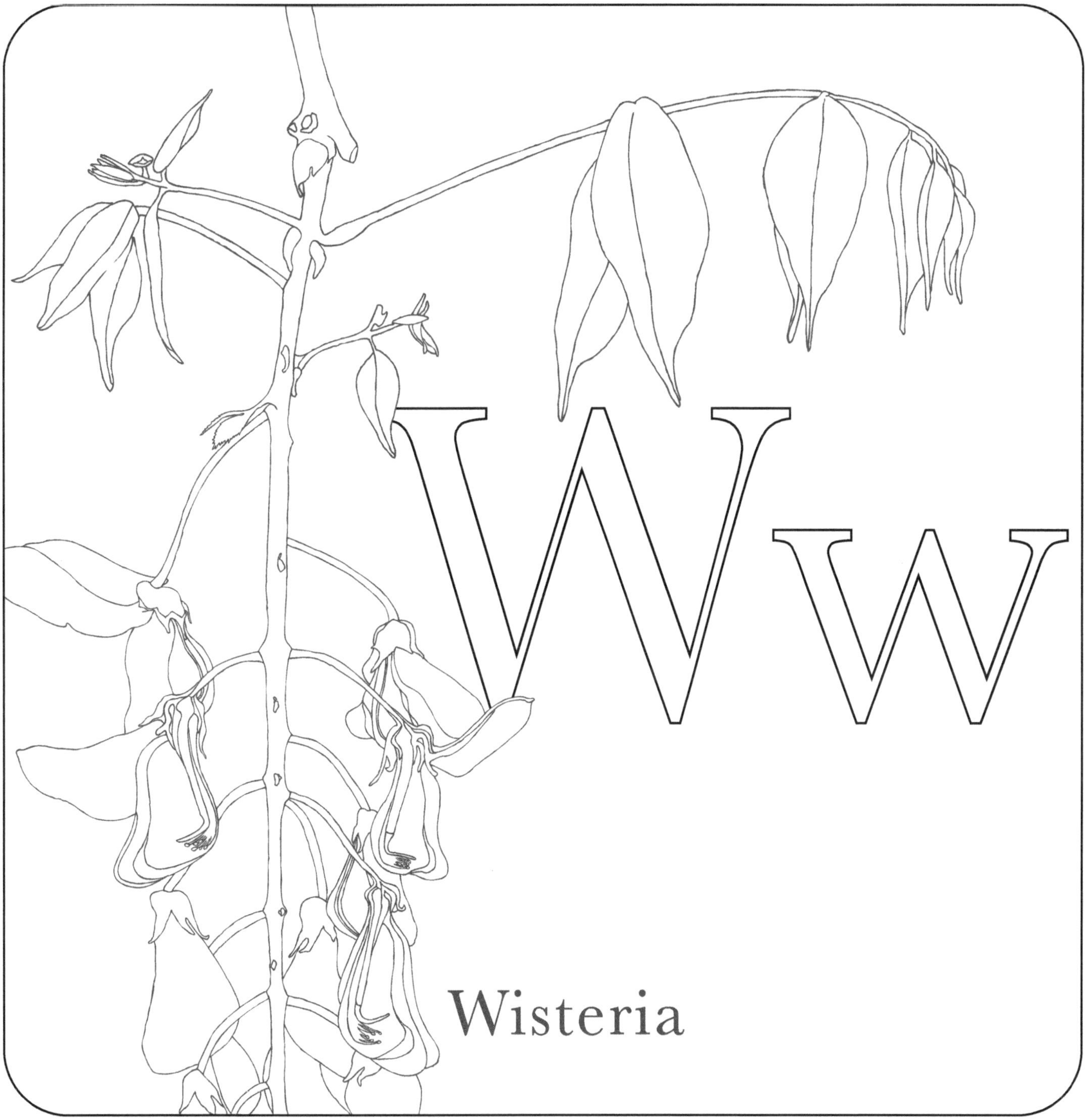
Ww
Wisteria

Xx

foXglove

Y y
Yucca

Zinnia

0
zero

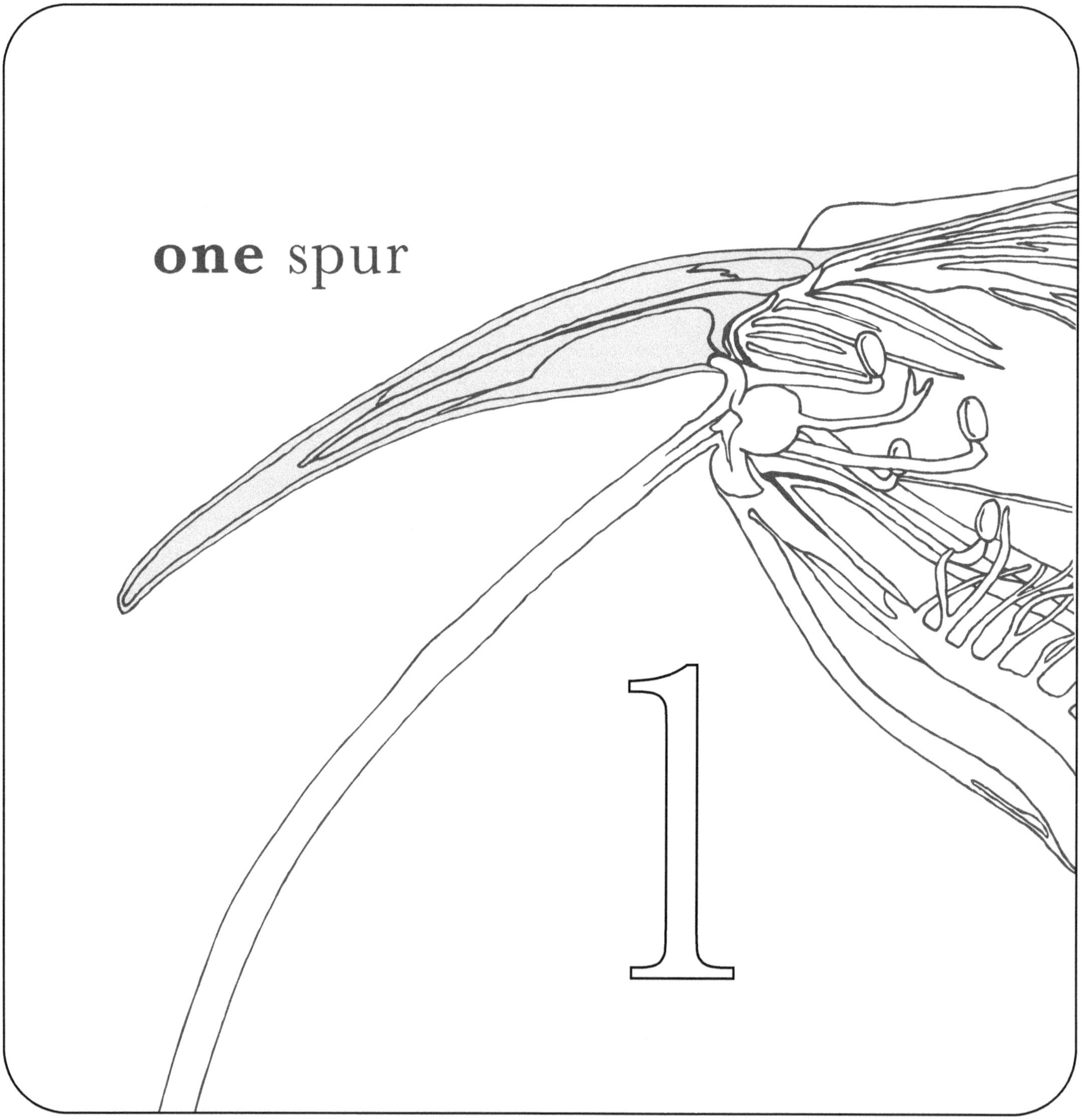
one spur
1

2
two thorns

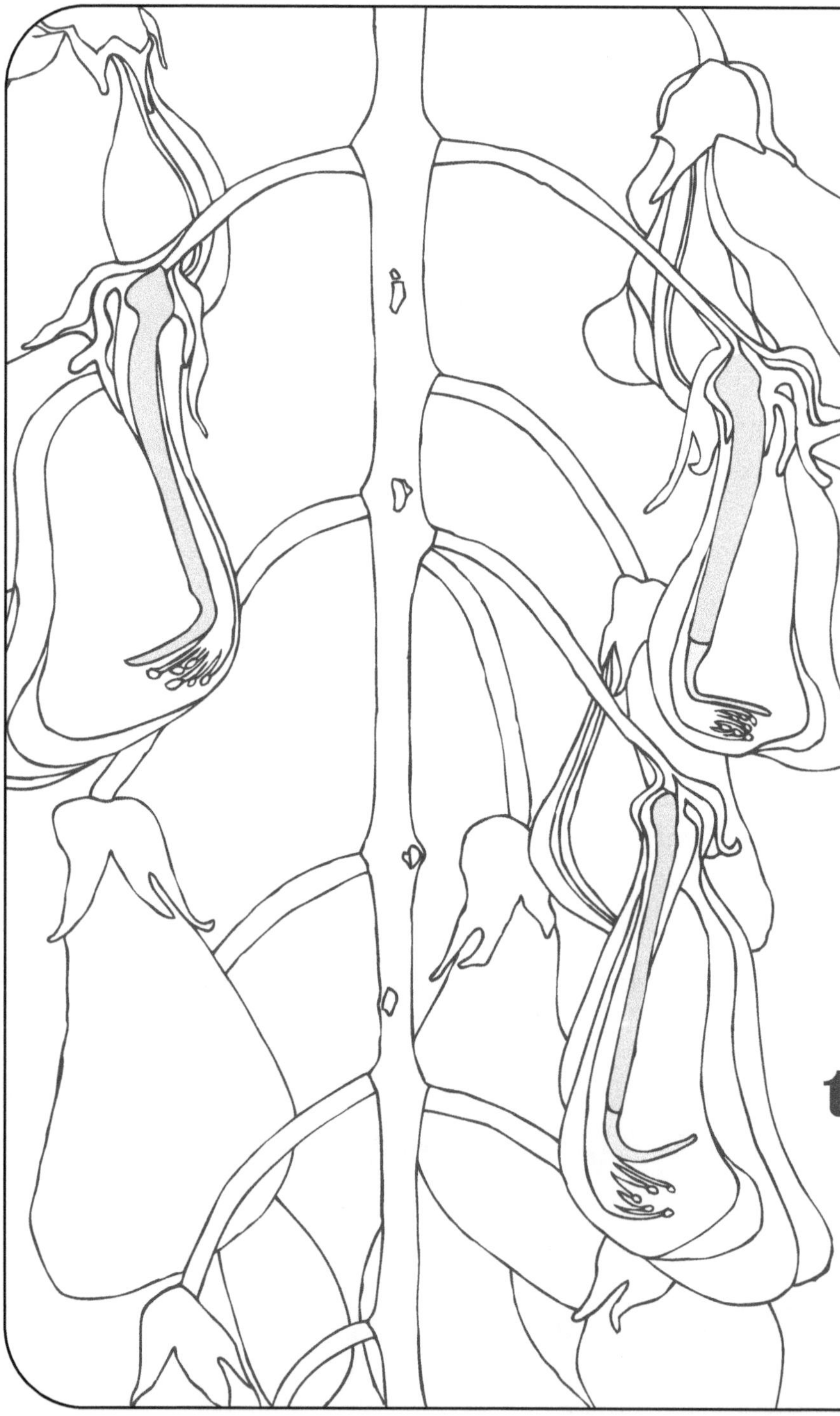

3

three pistils

four stamens
4

5

five tepals

6

six petals

7
seven sepals

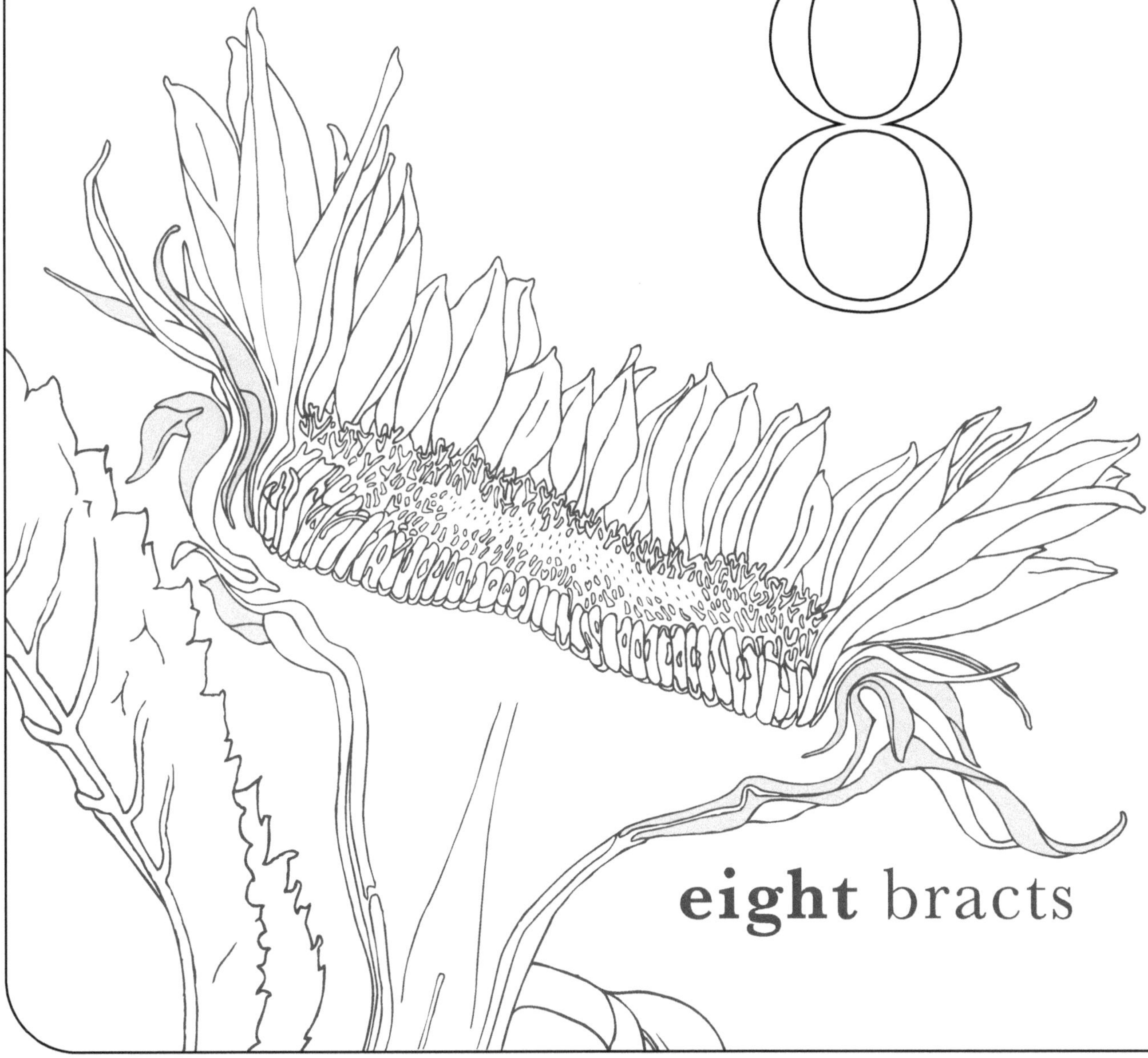

8

eight bracts

nine nectar guides

9

GLOSSARY

angiospermae - The division of the Plantae kingdom that contains all flowering plants. Angio means *covered* and sperm means *seed*. Angiosperms make *covered seeds* with their flowers.

angiospermophyta - phylum of the Plantae kingdom containing all flowering plants

annual - plant that completes its life cycle in one year

arrangement (leaf) - how the leaves are attached to the stem

atoms - tiny balls of energy that are the building blocks of cells

bilateral symmetry - only one set of halves is symmetrical

binomial nomenclature - a two-part naming system that Carl Linnaeus developed in the 1700s. Genus is capitalized. Species is lowercase.
Both are italicized: ***Genus species***

botany - the scientific study of plants

bract - modified leaf at the base of a flower or flower cluster that protects the buds and/or attracts pollinators. Bracts can be showy (the red petal-like parts of the poinsettia), dry and protective (sunflower), or papery (daffodil).

bulb - underground basal plate stems attached to scale leaves that store food for the plant

calyx - all the sepals of a single flower

capillary action - movement of water through very narrow tubes, allowing water to flow in opposition to gravity

cells - tiny building blocks that link together to build all living things

clade - a branch in the tree of life; in evolutionary terms, all members of a clade are descended from a common ancestor

common name - informal name that is widely used, often describing the physical or medicinal properties of the plant.

corolla - all the petals of a single flower

deciduous - plants that lose all their leaves annually

dicotyledoneae - archaic classification for a group of plants that produce seeds with two cotyledons (seed leaves)

DNA - **d**eoxyribo**n**eucleic **a**cid, a string of genetic information that determines the appearance and function of each living thing.

edible - safe for humans to eat

embryo (germ) - organism in the stage of the life after fertilization and into germination. This stage begins when a zygote divides into two cells and continues until it forms cotyledons (seed leaves), hypocotyl (primary stem), and radicle (primary root).

epicalyx - group of bracts arranged in a whorl located just under the sepals, which are located just under the petals (or tepals in monocots and magnoliidae)

eudicotyledoneae - a clade (branch) of flowering plants that produce seeds with two cotyledons (seed leaves)

evergreen - plants that retain leaves throughout the year

fieldwork - working in the field (real world) to gain experience and knowledge through observation and examination

florets - tiny flowers that make up a single flower. **Disc florets** can produce seeds and are in the center of the flower. **Ray florets** are much larger and usually have fused petals that look like individual petals around the edge of the flower. Sunflowers and echinacea flowers have florets.

flower - seed-producing part of a plant that has petals and sepals (or tepals), stamen/s, and pistil/s. Tiny flowers in a capitulum inflorescence are florets. Before the flower opens, it is called a bud.

fruit - the covering of an angiospermae seed

germination - the process of a seed embryo developing into a seedling

inflorescence - how flowers are arranged on their stem. A solitary inflorescence has a single flower on a stem (daffodils). A capitulum inflorescence has many florets side-by-side (sunflowers & echinacea).

interdependent- two or more things that are distinct in form and function yet dependent on one another for their existence

leaf - flat structures extending from the stem of the plant where photosynthesis takes place. Leaves are classified by venation (how veins are arranged), margin (edge), and shape (silhouette).

life cycle - the developmental changes of an organism throughout its life: germination, growth, reproduction (seeds), and death. *Cycle* means *circle*, so there is no beginning or end. For flowering plants, the process is: Seeds sprout and turn into seedlings. Seedlings become plants. Plants produce flowers that are pollinated, creating seeds. Not all plants produce seeds; some use a different process with spores.

magnoliidae - small, diverse clade of the Angiospermae Division containing plants that usually have pollen with a single pore and tepals like monocots, but often have two cotyledons and woody stems like the eudicots.

monocotyledoneae - clade (branch) of the Angiospermae Division that contains plants with one (mono) seed leaf (cotyledon)

nectar - sugary liquid produced by flowers to attract pollinators

nectar guides - markings on flowers that lead to the nectaries

nectaries - special plant parts that store nectar

organelle - interdependent (distinct yet dependent on one another) structures in a cell that perform specific tasks

organism - a living thing made up of interdependent parts

osmosis - flow of water from a high concentration area to a low concentration area though a semi-permeable membrane

ovary - protected area at the base of the pistil where **ovules** become seeds; the ovary becomes a fruit.

perennial - plant that lives beyond one year, often not mature enough to produce seeds for several years

perianth - all petals (corolla) and sepals (calyx) of a flower

petal - showy flower part that attracts pollinators

photosynthesis - process of converting energy from the sun into food. Special cell parts–chloroplasts–gather and store energy from the sun. Other plant parts can transport the energy packets to the entire plant. The basic formula is sunlight + carbon dioxide + water = sugar.

pistil - female part of the flower consistng of the stigma (sticky tip), style (long connector), and the ovary (where seeds develop)

pith - soft, spongy tissue that stores nutrients.

plant - multi-celled living orgaism that uses photosynthesis to create packets of energy (sugar)

poison - substance that causes physical harm or death

pollen - tiny grains attached to the anther of a stamen; contain DNA that can combine with an ovum's DNA to create new plants

pollination - transfer of pollen from the anther to the stigma of a pistil

pollinator - animal that aids in pollination

prickle - sharp points that emerge all along the stem but are not connected to the vascular system. (Roses have prickles, not thorns.)

propagation - increasing the number of a species through reproduction

protect - to keep safe from harm or injury

radial symmetry - identical wedges meeting at a central point; cuts through the center result symmetrical halves

reproduction - creation of new living organisms using cells and genetic material from one or two existing organisms

rhizome - specialized stem that grows at or below ground, sending new roots down and new leaves up.

root - part of the plant, usually underground, that collects and transports nutrients and water, anchors the plant, and often stores nutrients and food (carbohydrates). Roots can also float in water (ie. water lilies), and grab on to other plants (ie. orchids). Orchid roots can make food for the plant (photosynthesis) like leaves.

scientific - using the methods and principles of science

seed - unit that contains a plant embryo and is protected by a seed coat

sepal - petal-like or leaf-like part that surrounds and protects the petals as they develop

sexual reproduction - generation of new plants from the sexual parts (stamen, pistil) of existing plants

shape (leaf) - silhouette of the leaf

spathe - large bract around a cluster of tiny flowers, often mistaken for a single petal (ie. peace lily)

stamen - male part of the flower that includes pollen on the anther (where pollen forms) and a filament (long attachment to the flower).

stem - part of the plant that provides structure and transports water, nutrients, and sugars up to the leaves

symmetrical - two halves of a bisected object having the same parts placed in the same places on both sides

tepals - petals and sepals that look and feel alike

thorns - single sharp points that emerge at nodes along the stem. Quince has thorns.

vegetative reproduction - generation of new plants from the vegetative material (roots, stems, leaves) of existing plants

venation (leaves) - how veins branch within the leaf

whorl - group of plant parts (leaves, petals, stamens, etc.) that radiate from a common point; there must be more than two parts: two leaves originating from the same point would be *opposite*, not a whorl

www.ingramcontent.com/pod-product-compliance
Lightning Source LLC
LaVergne TN
LVHW070901210926
848482LV00016B/990